GRAPHING CALCULATOR MANUAL

DAPHNE BELL

Motlow State Community College

CALCULUS AND ITS APPLICATIONS

NINTH EDITION

D0140643

Marvin L. Bittinger

Indiana University Purdue University Indianapolis

David J. Ellenbogen

Community College of Vermont

PEARSON

Addison Wesley

Boston San Francisco New York
London Toronto Sydney Tokyo Singapore Madrid
Mexico City Munich Paris Cape Town Hong Kong Montreal

Reproduced by Pearson Addison-Wesley from electronic files supplied by the author.

Copyright © 2008 Pearson Education, Inc.
Publishing as Pearson Addison-Wesley, 75 Arlington Street, Boston, MA 02116.

ISBN-13: 978-0-321-45582-6
ISBN-10: 0-321-45582-7

1 2 3 4 5 6 OPM 10 09 08 07

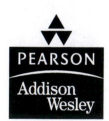

Contents

The TI-83 Plus and TI-84 Plus

Graphing Calculators

Chapter R
Functions, Graphs, and Models

The TI-83 Plus and TI-84 Plus families of calculators operate essentially the same way. Their keystrokes and screens are almost identical. When there is a difference, the TI-83 Plus screen will be on the left and the TI-84 Plus screen will be on the right unless otherwise indicated. Any difference in keystrokes will be noted.

Most of the applications that come pre-loaded on the TI-84 Plus are available for the TI-83 Plus as a free download from Texas Instruments on their web site—www.education.ti.com. User manuals are also available.

GETTING STARTED

Press ON to turn on the calculator. (ON is the key at the bottom left-hand corner of the keypad.) You should see a blinking rectangle, or cursor, on the screen. If you do not see the cursor, try adjusting the display contrast. To do this, first press 2nd. (2nd is in the upper left portion of the keypad. It is the yellow key on the TI-83 Plus and the blue key on the TI-84 Plus.) Then press and hold ▲ to increase the contrast or ▼ to decrease the contrast. If the contrast needs to be adjusted further after the first adjustment, press 2nd and then hold ▲ or ▼ to increase or decrease the contrast, respectively.

Press MODE to display the MODE settings. Initially you should select the settings on the left side of the display, as seen on the screens below.

To change a setting on the mode screen use ▲ or ▼ to move the cursor to the line of that setting. Then use ▶ or ◀ to move the blinking cursor to the desired setting and press ENTER. Press CLEAR or 2nd [QUIT] to leave the mode screen. ([QUIT] is the second function associated with the MODE key.) In general, second functions are written in yellow above the keys on the keypad of the TI-83 Plus and in blue above the keys on the keypad of the TI-84 Plus. Both calculators also have an ALPHA key. It is the green key directly below the 2nd key. When it is pressed, each of the green alphabetic characters written above the keys can be accessed on the calculator.

Before proceeding read the Getting Started section of the Texas Instruments Guidebook.

SETTING THE VIEWING WINDOW

Section R.1, Technology Connection, page 7   (Page numbers refer to pages in the textbook.)

The viewing window is the portion of the coordinate plane that appears on the graphing calculator's screen. It is defined by the minimum and maximum values of x and y: Xmin, Xmax, Ymin, and Ymax. The notation

[Xmin, Xmax, Ymin, Ymax] is used in the text to represent these window settings. For example, [-12, 12, -8, 8] denotes a window that displays the portion of the x-axis from -12 to 12 and the portion of the y-axis from -8 to 8. In addition, the distance between tick marks on the axes is defined by the settings Xscl and Yscl. In this manual, Xscl and Yscl will be assumed to be 1 unless noted otherwise. The setting Xres sets the pixel resolution. We usually select Xres = 1. The window corresponding to the settings [-20, 30, -12, 20], Xscl = 5, Yscl = 2, Xres = 1, is shown below.

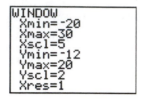 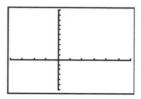

Press the WINDOW key on the top row of the keypad to display the current window settings on your calculator. The standard viewing window is shown below.

Equations are entered on the Y =, or equation-editor, screen. Press Y= to access this screen. If any plot (Plot 1, Plot 2, or Plot 3) is turned on (highlighted), turn it off by using the arrow keys to move the blinking cursor

To change a setting, position the cursor to the right of the setting you wish to change and enter the new value. For example, to change from the standard settings to [-20, 30, -12, 20], Xscl = 5, Yscl = 2, on the window screen press (-) 2 0 ENTER 3 0 ENTER 5 ENTER (-) 1 2 ENTER 2 0 ENTER 2 ENTER. You must use the (-) key on the bottom row of the keypad rather than the − key on the right-hand side of the keypad to enter a negative number. (-) represents "the opposite of" or "the additive inverse of" a number, whereas − is the key used for subtraction. The down-arrow, ▼, may be used instead of ENTER after typing each window setting. To see the window shown above, press the GRAPH key located in the upper right corner of the keypad.

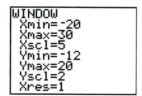 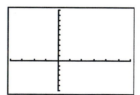

QUICK TIP: To return quickly to the standard viewing window [-10, 10, -10, 10], Xscl = 1, Yscl = 1, press ZOOM 6.

GRAPHS

After entering an equation and setting a viewing window, you can view the graph of an equation.

Section R.1, Technology Connection, page 7   Graph $y = x^3 - 5x + 1$ using a graphing calculator.

Equations are entered on the Y =, or equation-editor, screen. Press Y= to access this screen. If any plot (Plot 1, Plot 2, or Plot 3) is turned on (highlighted), turn it off by using the arrow keys to move the blinking cursor over the plot name and press $\boxed{\text{ENTER}}$. If there is currently an expression displayed for Y_1, clear it by positioning the cursor to the right of "$Y_1 =$" and pressing $\boxed{\text{CLEAR}}$. Remove all other expressions that appear in the equation-editor screen by pressing $\boxed{\blacktriangledown}$ and $\boxed{\text{CLEAR}}$. Then use $\boxed{\blacktriangle}$ and $\boxed{\blacktriangledown}$ to move the cursor to the right of "$Y_1 =$." Now press $\boxed{\text{X,T,}\Theta\text{,}n}$ $\boxed{\wedge}$ $\boxed{3}$ $\boxed{-}$ $\boxed{5}$ $\boxed{\text{X,T,}\Theta\text{,}n}$ $\boxed{+}$ $\boxed{1}$ to enter the right-hand side of the equation on the equation-editor screen.

The standard viewing window, [-10, 10, -10, 10], is a good choice for this graph. Either enter these dimensions, [-10, 10, -10, 10], on the window screen and then press $\boxed{\text{GRAPH}}$ or simply press $\boxed{\text{ZOOM}}$ $\boxed{6}$ to select the standard viewing window and see the graph.

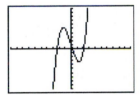

You can edit your entry if necessary. If, for instance, you pressed $\boxed{6}$ instead of $\boxed{5}$, when you put the equation into Y_1, you can use the $\boxed{\blacktriangleleft}$ key to move the cursor to 6 and then press $\boxed{5}$ to overwrite it. If you forgot to type the plus sign, move the cursor to the 1, and then press $\boxed{\text{2nd}}$ $\boxed{\text{INS}}$ $\boxed{+}$ to insert the plus sign before the 1. ($\boxed{\text{INS}}$ is the second function associated with the $\boxed{\text{DEL}}$ key.) You can continue to insert symbols immediately after the first insertion without pressing $\boxed{\text{2nd}}$ $\boxed{\text{INS}}$ again. If you typed 52 instead of 5, move the cursor to the 2 and press $\boxed{\text{DEL}}$ to delete the 2.

An equation must be solved for y before it can be graphed on any of the calculators in the TI-83 Plus or TI-84 Plus families.

Section R.1, Technology Connection, page 8; Example 2, page 5 To graph $3x + 5y = 10$, first solve for y,

obtaining $y = \dfrac{-3x + 10}{5}$. Then press $\boxed{\text{Y=}}$ and clear any expressions that currently appear. Position the cursor to the

right of "$Y_1 =$." Now press $\boxed{(}$ $\boxed{(-)}$ $\boxed{3}$ $\boxed{\text{X,T,}\Theta\text{,}n}$ $\boxed{+}$ $\boxed{1}$ $\boxed{0}$ $\boxed{)}$ $\boxed{\div}$ $\boxed{5}$ to enter the right-hand side of the equation. Without

the parentheses, the expression $-3x + \dfrac{10}{5}$ would have been entered.

Select a viewing window and then press $\boxed{\text{GRAPH}}$ to display the graph. You may change the viewing window as desired to reveal more or less of the graph. The graph is shown here using the standard viewing window.

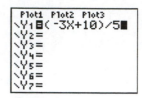   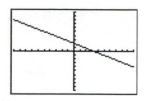

Section R.1, Technology Connection, page 8; Example 4, page 6 To graph $x = y^2$, first solve the equation for y, obtaining $y = \pm\sqrt{x}$. To see the entire graph of $x = y^2$ we must graph $Y_1 = \sqrt{x}$ and $Y_2 = -\sqrt{x}$ on the same screen. Press [Y=] and clear any expressions that currently appear. With the cursor to the right of "$Y_1 =$" press [2nd] [√] [X,T,Θ,n] [)]. ([√] is the second function associated with the [x^2] key.) A left parenthesis appears along with the radical symbol, so a separate keystroke is not necessary to introduce it.

Here are three ways to enter $Y_2 = -\sqrt{x}$.

The first method is to enter the expression $-\sqrt{x}$ directly. Use the down-arrow key or press [ENTER] to move the cursor to the right of "$Y_2 =$." Then press [(-)] [2nd] [√] [X,T,Θ,n] [)].

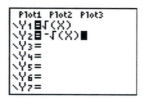

The second method of entering $Y_2 = -\sqrt{x}$ is based on the fact that $-\sqrt{x}$ is the additive inverse of $\sqrt{x}$. In other words, $Y_2 = -Y_1$. Move the cursor to the right of "$Y_2 =$" and press [(-)] [VARS] [▶] to go to the Y-VARIABLES submenu of the VARIABLES menu. Then press [1] [1] or [ENTER] [ENTER] to select "Y_1."

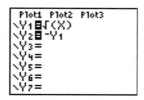

The third method of entering $Y_2 = -\sqrt{x}$ involves using the RECALL feature. Move the cursor to the right of "$Y_2 =$" and press [(-)] [2nd] [RCL]. ([RCL] is the second function associated with the [STO▶] key.) To recall $Y_2 = -\sqrt{x}$ press [VARS] [▶] to go the Y-VARIABLES submenu of the VARIABLES menu. Then press [1] [1] or [ENTER] [ENTER] to select "Y_1." Press [ENTER] again and $\sqrt{x}$ is pasted on the screen. This method is particularly useful if the expression to be copied is lengthy.

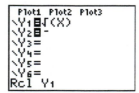   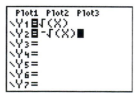

Once both the equations are on the equation-editor screen, select a viewing window and press GRAPH to display the graph. The window shown here is [-2, 10, -5, 5].

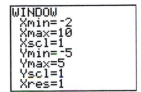

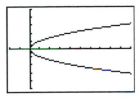

The top half is the graph of Y_1, the bottom half is the graph of Y_2, and together they yield the graph of $x = y^2$.

THE FINANCE APPLICATION

Both the TI-83 Plus and TI-84 Plus families of calculators have an APPS key. This is short for applications. Some of the applications are preloaded on the calculators and others can be downloaded from the TI web site. The FINANCE APPLICATION is on all the calculators in both families.

NOTE: If you are using an older model, such as the TI-83, the FINANCE APPLICATION is the second function on the x^{-1} key.

Section R.1, Example 7, page 11 Suppose that $1000 is invested at 8%, compounded quarterly. How much is in the account at the end of 3 years?

Press the APPS key. Depending upon which applications you have you may see a different screen than the one below.

Choose "1:Finance" by pressing 1 or ENTER. Press 1 or ENTER to choose "1:TVM Solver" (Time-Value-Money Solver). Use the arrow keys to navigate the screen.

N is the number of compoundings per year multiplied by the number of years. Here **N** will be 12. **I%** is 8. **PV** stands for Present Value and is $1000. It is entered as a negative value because it is considered outflow (investment). **PMT** stands for payment. This is 0 in this problem. **FV** stands for Future Value. Future Value is the unknown in this problem. **P/Y** and **C/Y** stand for payments year and compoundings per year, respectively. These numbers are both 4 because the interest is going back into the account quarterly until the 3-year period is over.

Once your screen looks like the one below you are ready to solve the problem.

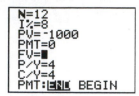

Move the cursor to the **FV** prompt as in the screen on the left below and press [ALPHA] [SOLVE]. ([SOLVE] is the alphabetic function associated with the [ENTER] key.)

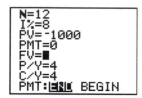   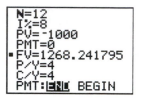

The answer is $1268.24.

To leave the screen, press [2nd] [QUIT].

THE TABLE FEATURE

For an equation entered in the equation-editor screen, a TABLE of x- and y-values can be displayed.

Section R.2, Technology Connection, page 17 Create a table of ordered pairs for the function $f(x) = x^3 - 5x + 1$.

Enter the function as $Y_1 = x^3 - 5x + 1$ as described on page 5 of this manual. Once the equation is entered, press [2nd] [TBLSET] to display the table set-up screen. ([TBLSET] is the second function associated with the [WINDOW] key.) A starting value for x can be chosen along with an increment for the x-value. To select a starting value of 0 and an increment of 1, press [0] and [ENTER] or [▼] and then press [1]. The "Indpnt:" and "Depend:" settings should both be "Auto." If either is not, use the [▼] to position the blinking cursor over "Auto" and then press [ENTER]. To display the table, press [2nd] [TABLE]. ([TABLE] is the second function associated with the [GRAPH] key.)

      (third screen)

Use the [▲] and [▼] keys to scroll through the table. For example, by using [▼] to scroll down we can see that $Y_1 = 1277$ when $x = 11$. Using [▲] to scroll up, observe that $Y_1 = -11$ when $x = -3$.

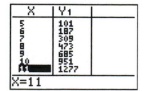

GRAPHS and FUNCTION VALUES

Section R.2, Technology Connection, page 19   There are several ways to evaluate a function using a graphing calculator. Three of them are described here. Given the function $f(x) = 2x^2 + x$, we will find $f(-2)$.

First press $\boxed{Y=}$ and enter the function as $Y_1 = 2x^2 + x$. Now we will find $f(-2)$ using the TABLE feature. Press $\boxed{2nd}$ $\boxed{\text{TBLSET}}$. Move the cursor to "Indpnt:" and highlight "Ask." Press $\boxed{\text{ENTER}}$. When "Ask" is highlighted, you supply the x-values and the graphing calculator returns the corresponding y-values. The settings for "TblStart" and "$\triangle$ Tbl" are irrelevant in this mode. Press $\boxed{2nd}$ $\boxed{\text{TABLE}}$ and find $f(-2)$ by pressing $\boxed{(-)}$ $\boxed{2}$ $\boxed{\text{ENTER}}$. We see that $Y_1 = 6$ when $x = -2$, so $f(-2) = 6$.

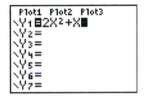  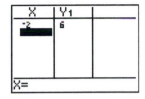

We can also use the VALUE feature from the CALCULATE menu to find $f(-2)$. To do this, graph $f(x) = 2x^2 + x$ in a window that includes the x-value, -2. We will use the standard viewing window. Press $\boxed{\text{ZOOM}}$ $\boxed{6}$ to graph. Press $\boxed{2nd}$ $\boxed{\text{CALC}}$ $\boxed{1}$ or $\boxed{2nd}$ $\boxed{\text{CALC}}$ $\boxed{\text{ENTER}}$ to select "1:value" from the CALCULATE menu. ($\boxed{\text{CALC}}$ is the second function associated with the $\boxed{\text{TRACE}}$ key.) Now press $\boxed{(-)}$ $\boxed{2}$ $\boxed{\text{ENTER}}$ to see "$x = -2$" and "$y = 6$" at the bottom of the screen. Again we see that $f(-2) = 6$.

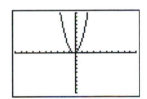 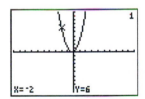

The third method for finding $f(-2)$ uses function notation on the home screen of the calculator. With $Y_1 = 2x^2 + x$ entered on the equation-editor screen, go to the home screen by pressing $\boxed{2nd}$ $\boxed{\text{QUIT}}$. ($\boxed{\text{QUIT}}$ is the second function associated with the $\boxed{\text{MODE}}$ key.) Now enter $Y_1(-2)$ by pressing $\boxed{\text{VARS}}$ $\boxed{\blacktriangleright}$ to see the y-variables. Press $\boxed{1}$ to choose "1:Function." Press $\boxed{1}$ to choose "1:Y_1." Then press $\boxed{(}$ $\boxed{(-)}$ $\boxed{2}$ $\boxed{)}$ $\boxed{\text{ENTER}}$. Again we see that $f(-2) = 6$.

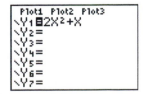 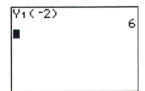

THE TRACE FEATURE

The TRACE feature can be used to display the coordinates of points on a graph.

Section R.2, Technology Connection, page 19   Use the function $f(x) = 2x^2 + x$ graphed in the standard viewing

window. Press $\boxed{Y=}$ and put the function on the equation-editor screen. Then press $\boxed{ZOOM}$ $\boxed{6}$.

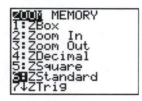   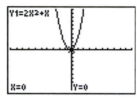   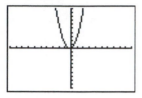

Press $\boxed{TRACE}$ and a blinking cursor appears on the graph at the x-value that is the x-coordinate of the

midpoint of the x-axis. The y-coordinate associated with this x-coordinate is also displayed.

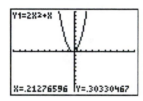

Press $\boxed{\blacktriangleright}$ and another ordered pair will be displayed at the bottom of the screen.

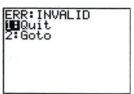

Press $\boxed{TRACE}$ $\boxed{3}$ $\boxed{ENTER}$ to see that when $x = 3$ the associated y-coordinate is 21. Even though 21 is larger

than the maximum y-value in the window, it is still displayed at the bottom of the screen.

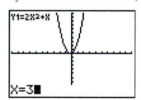   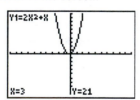

If you enter an x-value that is not between the minimum and maximum window values for x, an error will

result. Try it!

Press $\boxed{TRACE}$ $\boxed{2}$ $\boxed{0}$ $\boxed{ENTER}$ and you will see an error results. Press $\boxed{1}$ or $\boxed{ENTER}$ to select "1: Quit."

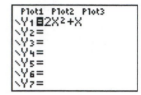

GRAPHING FUNCTIONS DEFINED PIECEWISE; DOT MODE

Section R.2, Technology Connection, page 23; Example 9, page 22   Graph: $f(x) = \begin{cases} 4 & for \ \ x \le 0 \\ 3 - x^2 & for \ \ 0 < x \le 2 \\ 2x - 6 & for \ \ x > 2 \end{cases}$.

We will enter the function using inequality symbols from the TEST menu. Press [Y=] to go to the equation-editor screen. Clear any entries that are present. Position the cursor to the right of "Y₁ =" and press [( ] [4] [ )] [( ]
[X,T,Θ,n] [2nd] [TEST] [6] [0] [ )] [+] [( ] [3] [−] [X,T,Θ,n] [x²] [ )] [( ] [0] [2nd] [TEST] [5] [X,T,Θ,n] [ )] [( ] [X,T,Θ,n] [2nd] [TEST] [6] [2] [ )]
[+] [( ] [2] [X,T,Θ,n] [−] [6] [ )] [( ] [X,T,Θ,n] [2nd] [TEST] [3] [2] [ )]. ([TEST] is the second function associated with the [MATH]
key.) The keystrokes [2nd] [TEST] [6] open the TEST menu and select item "6: ≤ " from that menu. Similarly,
[2nd] [TEST] [5] and [2nd] [TEST] [3] open the TEST menu and select "5: <" and "3: >," respectively.

When graphing a piece-wise defined function, put the calculator in DOT mode. In CONNECTED mode
the calculator will try to connect pieces of the graph that should not be connected. DOT mode can be selected in
two ways. One is to press [MODE], move the cursor over "Dot," and press [ENTER]. DOT mode can also be selected on
the equation-editor screen by positioning the cursor over the graph style icon to the left of "Y₁ =" and pressing
[ENTER] six times until the dotted graph style icon appears.

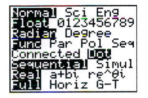      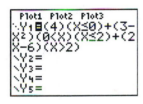

Enter the window dimensions and press [GRAPH] to display the graph of the function. It is shown here in the
window [-5, 5, -3, 6].

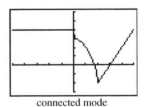   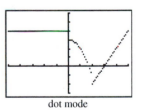

connected mode   dot mode

When the equation in Y₁ is cleared, the graph style will automatically default back to CONNECTED mode.

SQUARING THE VIEWING WINDOW

Section R.4, Technology Connection, page 40   In the standard viewing window, the distance between tick marks
on the y-axis is about $\dfrac{2}{3}$ the distance between tick marks on the x-axis. It is often desirable to choose window
dimensions for which these distances are the same, creating a "square" window. Any window in which the ratio of

the length of the y-axis to the length of the x-axis is $\dfrac{2}{3}$ will produce this effect. This can be accomplished by

selecting dimensions for which Ymax − Ymin = $\dfrac{2}{3}$(Xmax − Xmin).

The standard viewing window is shown on the left below and the square window [-6, 6, -4, 4] is shown on the right. Observe that the distance between tick marks appears to be the same on both axes in the square window.

 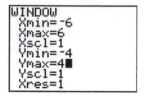 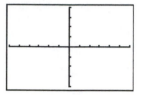

The window can also be squared by pressing ZOOM 5 to select "5:ZSquare" from the ZOOM menu.

Starting with the standard viewing window and pressing ZOOM 5 produces the dimensions and the window shown below.

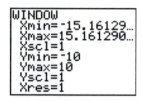 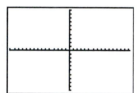

THE INTERSECT FEATURE

Section R.4, Example 10, page 49 *Business: Profit-and-Loss Analysis* When a business sells an item, it receives the price paid by the consumer. This is normally greater than the cost to the business of producing the item.

(a)   The total revenue that a business receives is the product of the number of items sold and the price paid per item. Thus, if Raggs, Ltd. sells x suits at \$80 per suit, the total revenue, $R(x)$, is given by $R(x) = 80x$.

The cost equation, $C(x) = 20x + 10{,}000$, is found in Example 9 on page 48 in your text book. Graph both $R(x)$ and $C(x)$ on the same set of axes.

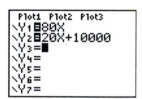

The window must be modified to fit the problem. Certainly it makes no sense to have a negative number of products; therefore, since the x-axis is the item axis, the minimum value should be 0. The y-axis is the money axis and can certainly be negative, but for this first graph is not. The window settings and the graphs of the revenue and cost equations are on the next page.

   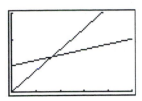

(b)   The total profit a business makes is the money that is left after all costs have been subtracted from

the total revenue. If $P(x)$ represents the profits when x items are produced and sold, we have $P(x) = R(x) - C(x)$.

Determine $P(x)$ and draw its graph using the same set of axes as those used for the graph in part (a).

The revenue function is in Y_1 and the cost function is in Y_2, therefore $P(x) = Y_1 - Y_2$. Position the cursor

to the right of "$Y_3 =$" and press [VARS] [▶] [1] [1] [−] [VARS] [▶] [1] [2]. To allow for the profit to be negative, the window

has been modified. All three graphs are displayed on the same set of axes.

   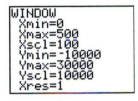   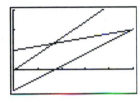

(c)   The company will break even at the value of x for which $P(x) = 0$ (that is, no profit and no loss).

This is the point at which $R(x) = C(x)$. Find the break-even point.

Use the INTERSECT feature of the CALCULATE menu to find the break-even point. With the graph

displayed, press [2nd] [CALC] [5] to choose "5:Intersect." ([CALC] is the second function associated with the [TRACE] key.)

The calculator will pose three questions.

The query "First Curve?" appears at the bottom of the screen. The blinking cursor is positioned on the

graph of Y_1 as in the screen below on the left. Notice the upper left corner of this screen displays the equation of the

curve the cursor is indicating. Press [ENTER] to indicate that this is the first curve involved in the intersection.

Next the query "Second Curve?" appears at the bottom of the screen. The blinking cursor is now

positioned on the graph of Y_2 as in the screen below on the right. Notice the upper left corner of this screen displays

the equation of the curve the cursor is indicating. Press [ENTER] to indicate that this is the second curve. We identify

the curves for the calculator since we have three graphs on the screen.

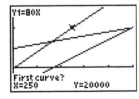   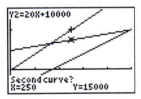

After we identify the second curve, the query "Guess?" appears at the bottom of the screen. Since there is

only one point of intersection, press [ENTER] and the point of intersection is displayed. This is the break-even point.

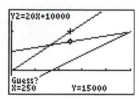   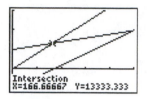

After finding the break-even point press ⬇ and you will see the screen below. This screen shows the value of the profit function when the *x*-coordinate of the break-even, or equilibrium, point is substituted into it. Notice the profit is $0 at the break-even point.

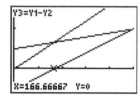

The company breaks even when 167 items are produced and sold.

THE MAXIMUM FEATURE

The MAXIMUM feature in the CALCULATE menu can be used to find the vertex of a parabola on the graph screen.

Section R.5, Example 2, page 57   Graph $f(x) = -2x^2 + 10x - 7$.

Graph the function as $Y_1 = -2x^2 + 10x - 7$ on the standard viewing window. The screen below on the left should appear. Press [2nd] [CALC] to go to the CALCULATE menu. To put the graph on graph paper, you need to find a few points. The most important point is the vertex. Since this parabola is concave down, the vertex is its maximum point. Press [4] to choose "4:maximum." In the bottom left corner of the screen "Left Bound?" appears. Move the cursor using the left- and/or right-arrow keys until the cursor is in a location on the curve that is to the left of the vertex then press [ENTER]. When you press [ENTER] a marker appears on the screen to mark the left boundary. "Right Bound?" now appears on the bottom of the screen.

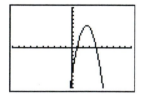   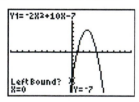   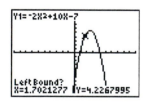   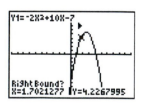

Move the cursor to the right of the vertex and press [ENTER]. Another marker appears, this one marks the right boundary. "Guess?" appears on the bottom of the screen. Move the cursor near the vertex of the parabola and press [ENTER]. The coordinates of the vertex are displayed.

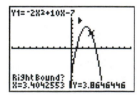   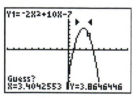   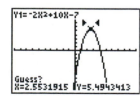   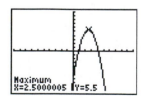

Sometimes the calculator does not give the exact answer. You must realize that 2.5000005 is really just 2.5. It is worth noting that the *x*- and *y*-values are temporarily stored in memory. Go to the home screen by pressing 2nd [QUIT]. The *x*-coordinate is 2.5. Press 2 . 5 and turn it into a fraction by pressing MATH 1 to choose "1:Frac." Press ENTER. To find the *y*-value press ALPHA Y (Y is the alphabetic character associated with the 1 key.) Press MATH 1 to choose "1:Frac" and press ENTER to turn it into a fraction.

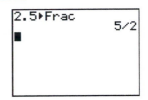

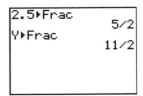

To complete the graph you need more points than just the vertex. Use the TABLE feature to find more points. Press 2nd [TBLSET] to display the table set-up screen. ([TBLSET] is the second function associated with the WINDOW key.) A starting value of *x* can be chosen along with an increment for the *x*-value. To select a starting value of 0 and an increment of 1, press 0 and ENTER or ▼ and then press 1. The "Indpnt:" and "Depend:" settings should both be "Auto." If either is not, use the ▼ to position the blinking cursor over "Auto" and press ENTER. To display the table, press 2nd [TABLE]. ([TABLE] is the second function associated with the GRAPH key.)

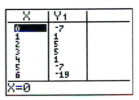

Notice the vertex is not on the table since the increment is 1, but you can tell that the *x*-coordinate of the vertex is half way between the *x*-values 2 and 3.

THE POLYSMLT APPLICATION—Polynomial Root-Finder

Both the TI-83 Plus and TI-84 Plus families of calculators have an APPS key. This is short for Applications. The POLYSMLT application will solve polynomial equations and will also solve systems of simultaneous equations. It is pre-loaded on some of the calculator models and is available to be downloaded free for others. If you do not have it on your calculator, you can download it at www.education.ti.com. The following example will be solved using the application and will be solved graphically using the INTERSECT feature on page 16 of this manual. It is solved algebraically in the text book.

Section R.5, Example 3, page 58   Solve: $3x^2 - 4x = 2$.

First write the equation in standard form $ax^2 + bx + c = 0$, and then determine *a*, *b*, and *c*:

$3x^2 - 4x - 2 = 0$, so $a = 3$, $b = -4$, and $c = -2$.

Press APPS use the up- and down-arrow keys to scroll until you find "PolySmlt," "PolySmlt2," or another abbreviation for "Polynomial Root-Finder and Simultaneous Equation Solver." The abbreviation will vary depending on the version of the application you have. The screen on the left is an older version than the one on the right. Move the selection cursor immediately to the left of the name of the application and press ENTER.

   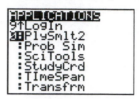

You may see a slightly different screen depending on the version you have. Press any key to get to the MAIN menu. Once in the MAIN menu select "1:PolyRoot Finder" by pressing $\boxed{1}$. Mark your screen as the third screen from the left is marked. (If you have an older version you may only have to give the degree and press ENTER before entering the coefficients.) Notice the choices along the bottom of the screen. Since "NEXT" is above the $\boxed{\text{GRAPH}}$ key press $\boxed{\text{GRAPH}}$ to go to the next screen. In the calculator a_2 is the coefficient of x^2, a_1 is the coefficient of x, and a_0 is the constant. Enter the coefficients as in the screen below on the right.

         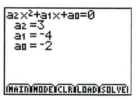

Notice the choices along the bottom of the calculator screen. Choose "SOLVE" by pressing the $\boxed{\text{GRAPH}}$ key. The decimal approximations of the answers are displayed even though "FRAC" was marked on the screen above. To exit the application, return to the MAIN menu by pressing $\boxed{\text{Y=}}$ then choose "6:QUIT POLYSMLT" by pressing $\boxed{6}$ or by highlighting the 6 and pressing ENTER.

THE INTERSECT FEATURE

We can use the INTERSECT feature from the CALCULATE menu to solve equations.

Section R.5, Example 3, page 58   Solve: $3x^2 - 4x = 2$.

Now we will solve this problem graphically. If we consider the left and right sides of the equation as two separate equations, the solutions to the equation will be the x-coordinates of the points of intersection.

Press $\boxed{\text{Y=}}$ clear any existing entries and enter $Y_1 = 3x^2 - 4x$ and $Y_2 = 2$. Press $\boxed{\text{ZOOM}}$ $\boxed{6}$ to display the graph on a standard viewing window.

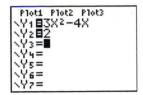 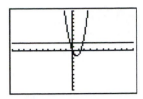

Use the INTERSECT feature from the CALCULATE menu to find the points of intersection. With the graph displayed, press [2nd] [CALC] [5] to choose "5:intersect." The calculator will pose three questions.

The query "First Curve?" appears at the bottom of the screen. The blinking cursor is positioned on the graph of Y_1 as in the screen below on the left. Notice the upper left corner of this screen displays the equation of the curve the cursor is indicating. Press [ENTER] to indicate that this is the first curve involved in the intersection.

Next the query "Second Curve?" appears at the bottom of the screen. The blinking cursor is now positioned on the graph of Y_2 as in the screen below on the right. Notice the upper left corner of this screen displays the equation of the curve the cursor is indicating. Press [ENTER] to indicate that this is the second curve. We identify the curves for the calculator since we could have as many as ten graphs on the screen at once.

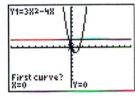 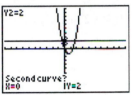

After we identify the second curve, the query "Guess?" appears at the bottom of the screen. Since the cursor is near one of the points of intersection, press [ENTER] and the point of intersection nearest the guess is displayed.

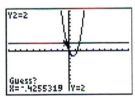 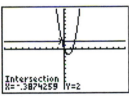

To find the second point of intersection, repeat the process. Make sure you move the cursor near the second point of intersection before you press [ENTER].

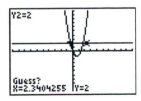 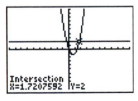

The two answers, rounded to three decimal places, are $x \approx -0.387$ or $x \approx 1.721$.

THE INTERSECT FEATURE

Another example of how the INTERSECT feature from the CALCULATE menu can be used to solve equations.

Section R.5, Technology Connection, page 59   Given, $f(x) = x^2 - 6x + 8$, find the solutions of $x^2 - 6x + 8 = 0$.

If we consider the left and right sides of the equation as two separate equations, the solutions to the equation will be the x-coordinates of the points of intersection.

Press Y= clear any existing entries and enter $Y_1 = x^2 - 6x + 8$ and $Y_2 = 0$. Press ZOOM 6 to display the graph on a standard viewing window. Remember that $Y_2 = 0$ is the graph of the x-axis, so you will not see another line since the axis is already on the screen.

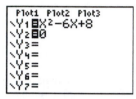   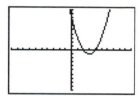

Use the INTERSECT feature from the CALCULATE menu to find the points of intersection. With the graph displayed, press 2nd [CALC] 5 to choose "5:intersect." The calculator will pose three questions.

The query "First Curve?" appears at the bottom of the screen. The blinking cursor is positioned on the graph of Y_1 as in the screen below on the left. Notice the upper left corner of this screen displays the equation of the curve the cursor is indicating. Press ENTER to indicate that this is the first curve involved in the intersection.

Next the query "Second Curve?" appears at the bottom of the screen. The blinking cursor is now positioned on the graph of Y_2 as in the screen below on the right. Notice the upper left corner of this screen displays the equation of the curve the cursor is indicating. Press ENTER to indicate that this is the second curve. Notice that the cursor is on the x-axis. Again, that is because $Y_2 = 0$ is the graph of the x-axis.

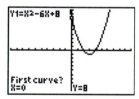   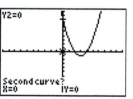

After we identify the second curve, the query "Guess?" appears at the bottom of the screen. Since the cursor is near one of the points of intersection, press ENTER and the point of intersection nearest the guess is displayed.

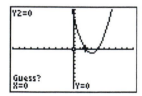   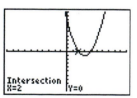

To find the second point of intersection, repeat the process. Make sure you move the cursor near the second point of intersection before you press ENTER.

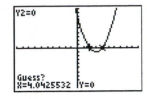 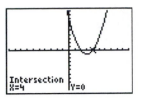

The two answers are $x = 4$ or $x = 2$.

THE INTERSECT FEATURE

Another example of how the INTERSECT feature from the CALCULATE menu can be used to solve equations.

Section R.5, Technology Connection, page 61   Solve the equation $x^3 = 3x + 1$.

If we consider the left and right sides of the equation as two separate equations, the solutions to the equation will be the x-coordinates of the points of intersection.

Press $\boxed{Y=}$ clear any existing entries and enter $Y_1 = x^3$ and $Y_2 = 3x + 1$. Now graph these equations in an appropriate window. One good choice is [-3, 3, -10, 10].

 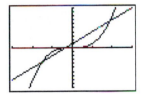

Use the INTERSECT feature from the CALCULATE menu to find the points of intersection. With the graph displayed, press $\boxed{2nd}$ $\boxed{CALC}$ $\boxed{5}$ to choose "5:intersect." The calculator will pose three questions.

The query "First Curve?" appears at the bottom of the screen. The blinking cursor is positioned on the graph of Y_1 as in the screen below on the left. Notice the upper left corner of this screen displays the equation of the curve the cursor is indicating. Press $\boxed{ENTER}$ to indicate that this is the first curve involved in the intersection.

Next the query "Second Curve?" appears at the bottom of the screen. The blinking cursor is now positioned on the graph of Y_2 as in the screen below on the right. Notice the upper left corner of this screen displays the equation of the curve the cursor is indicating. Press $\boxed{ENTER}$ to indicate that this is the second curve.

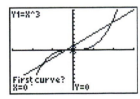 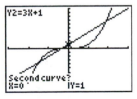

After we identify the second curve, the query "Guess?" appears at the bottom of the screen. Move the cursor near the leftmost point of intersection. Press $\boxed{ENTER}$. The point of intersection nearest the guess is displayed.

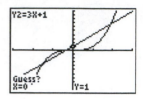 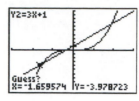 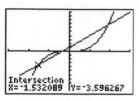

To find the other two points of intersection repeat the process for each point. Make sure you move the cursor near the point of intersection you are trying to find before you press ENTER.

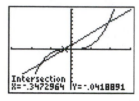 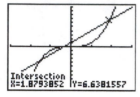

The three answers, rounded to two decimal places, are $x \approx -1.53$, $x \approx -0.35$ or $x \approx 1.88$.

THE ZERO FEATURE

Here is an example of how an equation can be solved using the ZERO feature from the CALCULATE menu. The equation to be solved must first be expressed in the form $f(x) = 0$. Many students prefer to use the INTERSECT feature, as described on pages 16, 18 and 19 of this manual, instead of the ZERO feature.

Section R.5, Technology Connection, page 61   Solve the equation $x^3 = 3x + 1$.

To use the ZERO feature, we must first have the equation in the form $f(x) = 0$. To do this, subtract $3x + 1$ on both sides of the equation to obtain an equivalent equation with 0 on one side. We have $x^3 - 3x - 1 = 0$. The solutions of the equation $x^3 = 3x + 1$ are the values of x for which the function $f(x) = x^3 - 3x - 1$ is equal to zero. We can use the ZERO feature to find these values, or zeros. In some texts, you will see the zeros referred to as the roots. These terms are interchangeable.

Go to the equation-editor screen by pressing Y=. Clear any existing entries and then enter $Y_1 = x^3 - 3x - 1$. Now graph the function in a viewing window that shows the x-intercepts clearly. One good choice is [-3, 3, -5, 8]. We see that the function has three zeros. They appear to be about -1.5, -0.5, and 2.

 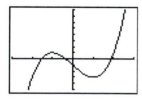

We will find the zero near -1.5 first. Press 2nd [CALC] 2 to select "2:zero" from the CALCULATE menu. We are prompted to select a left bound. This means that we must choose an x-value that is to the left of -1.5 on the x-axis. This can be done by using the left- and right-arrow keys to move the cursor to a point on the curve to the left of -1.5 or by keying in a value less than -1.5.

 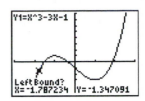

Once this is done press $\boxed{\text{ENTER}}$. Now we are prompted to select a right bound that is to the right of -1.5 on the *x*-axis. Again, this can be done by using the arrow keys to move the cursor to a point on the curve to the right of −1.5 or by keying in a value greater than -1.5.

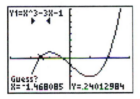

Press $\boxed{\text{ENTER}}$ again. Finally we are prompted to make a guess as to the value of the zero. Move the cursor to a point near the zero or key in a value.

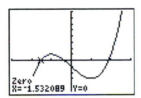

Press $\boxed{\text{ENTER}}$ a third time. We see that $y = 0$ when $x \approx -1.53$, so -1.53 is a zero of the function.

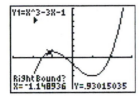

Select "2:zero" from the CALCULATE menu a second time to find the zero near -0.5 and a third time to find the zero near 2.

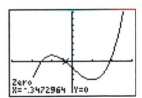 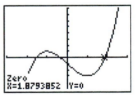

We see that the other two zeros are approximately -0.35 and 1.88.

ABSOLUTE-VALUE FUNCTIONS

We can use the absolute-value option to perform computations involving absolute value and to graph absolute-value functions.

Section R.5, Example 8, page 65   Graph $f(x) = |x|$.

The absolute-value function is found in the NUMBER submenu in the MATH menu and in the CATALOG. To graph $f(x) = |x|$ press Y= to go to the equation-editor screen and clear any existing entries. Now position the cursor to the right of "$Y_1 =$" and press MATH ▶ to access the NUMBER submenu of the MATH menu. Press ENTER or 1 to choose "1:abs(" from the menu. Then press X,T,Θ,n). The right parenthesis is necessary since the absolute-value function opened a parenthesis. To select "abs(" from the CATALOG, press 2nd [CATALOG] ENTER. ([CATALOG]is the second function associated with the 0 key.) In either case, choose an appropriate viewing window and press GRAPH to see the graph of the function. The graph is shown here in the standard viewing window.

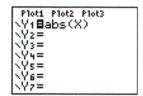
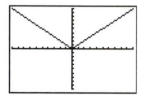

GRAPHING RADICAL FUNCTIONS

There are various ways to enter radical expressions on a graphing calculator.

Section R.5, Powers with Rational Exponents, page 66 We discussed entering an expression containing a square root on page 6 of this manual. We can use the cube root, "$\sqrt[3]{}($," function from the MATH menu to enter an expression containing a cube root, or we can raise the expression to the $\frac{1}{3}$ power. To enter $Y_1 = \sqrt[3]{x-2}$, for example, position the cursor immediately to the right of "$Y_1 =$" on the equation-editor screen and press MATH 4 to select "$4:\sqrt[3]{}($." Then press X,T,Θ,n − 2) to enter the radicand.

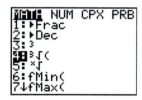
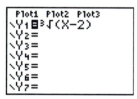

Press WINDOW and change your settings to the ones below. Then press GRAPH.

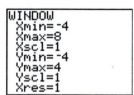
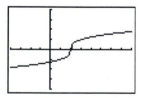

Remember that $\sqrt[3]{x-2} = (x-2)^{\frac{1}{3}}$. Return to the equation-editor screen by pressing Y= and with the cursor to the right of "$Y_2 =$" press (X,T,Θ,n − 2) ^ (1 ÷ 3). Before graphing this function, use the left-arrow key to move the cursor over the graph style icon to the left of "Y_2" and press ENTER 4 times until you see "⊹." This icon will be referred to as the path icon in this manual. Now press GRAPH to see what looks like a ball moving along the curve that was already graphed. This shows that the two ways of writing cube root are equivalent.

 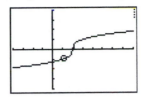

We could also compare the tables and see that the functions are the same. When we use the square root or cube root function, the calculator considers the entire expression to be the radicand, so it is not necessary to close the parentheses. We do so, nevertheless, for completeness.

Higher order radical expressions can be entered using the $\sqrt[x]{}$ function from the MATH menu. When we use this function, we must enclose the radicand in parentheses if it contains more than one term. Clear all existing entries on the equation-editor screen. To enter $Y_1 = \sqrt[4]{x-1}$, position the cursor to the right of "$Y_1 =$." Then press 4 to indicate that we are entering a fourth root. Next press MATH 5 to select "5: $\sqrt[x]{}$ " from the MATH menu and finally press (X,T,θ,n − 1) to enter the radicand. Note that the calculator does not automatically supply a left parenthesis as it does when a square root or a cube root is selected. The fourth root can also be written using a fractional exponent. To enter the alternate form of the function, position the cursor to the right of "$Y_2 =$" and press (X,T,θ,n − 1) ^ (1 ÷ 4). Prove to yourself they are the same by changing the graph style icon on one of them to path and graphing them or by examining the table. Here we examine the table. Press 2nd [TBLSET] and set your table as in the middle screen below. Then press 2nd GRAPH to see the table displayed. Notice the entries are the same. The ERROR message is because when taking an even root, the radicand's value can not be less than 0. This function does not exist when $x < 1$.

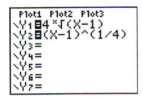  

LINEAR REGRESSION

We can use the LINEAR REGRESSION feature from the STATISTICAL CALCULATIONS menu to fit a linear equation to a set of data points.

Section R.6, Technology Connection, page 78; Example 2, page 77 The following table shows the average annual pay for a U. S. production worker.

Years, x, since 1996	1	2	3	4	5	6	7
Percentage increase since 1996, P	1.9	7.4	11.7	19.5	28.2	29.7	31.3

(a)   Make a scatter plot of the data and determine whether the data seem to fit a linear function.

We will enter the data as ordered pairs on the statistical list-editor screen. To clear existing lists press STAT 4 2nd [L1] , 2nd [L2] , 2nd [L3] , 2nd [L4] , 2nd [L5] , 2nd [L6] ENTER. (L_1 through L_6 are the second functions associated with the numeric keys 1 through 6.) The lists can also be cleared by first accessing the statistical list-editor screen by pressing STAT ENTER or STAT 1 to select "1:Edit" from the STATISTICAL menu. Then, for each list that contains entries, use the arrow keys to move the cursor to highlight the name of the list at the top of the column and press CLEAR ▼ or CLEAR ENTER.

Once the lists are cleared, we can enter the data points. We will enter the number of years since 1996 in L_1 and the percentage increase in L_2. Position the cursor at the top of column L_1 just below the L_1 heading. Press 1 ENTER. Continue typing the *x*-values 2 through 7, each followed by ENTER. The entries can be followed by ▼ rather than ENTER if desired. Press ▶ to move to the top of column L_2 just below the L_2 heading. Type the percentages 1.9, 7.4, and so on in succession, each followed by ENTER or ▼. Note that the coordinates of each point must be in the same position in both lists.

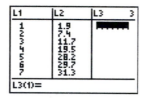

To plot the data points, we turn on the STATISTICAL PLOT feature. To access the statistical plot screen, press 2nd [STAT PLOT]. ([STAT PLOT] is the second function associated with the Y= key in the upper left-hand corner of the keypad.)

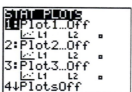

We will use "1:Plot 1." Since it is already highlighted, we access it by pressing ENTER or 1. The cursor should be positioned over "On" and it should be flashing. Press ENTER to turn on Plot 1. The entries Type, Xlist, and Ylist should be as shown below. The last item, Mark, allows us to choose a box, a cross, or a dot for each point. Here we have selected a box. To select Type and Mark, position the cursor over the appropriate selection and press ENTER to highlight the selection. Use the L_1 and L_2 keys (associated with the 1 and 2 numeric keys) to select Xlist and Ylist.

The plot can also be turned on from the equation-editor screen. Press $\boxed{Y=}$ to go to the equation-editor screen. Then assuming Plot 1 has not yet been turned on (is not highlighted) and that the desired settings are currently entered for Plot 1 on the statistical plot screen position the cursor over "Plot 1" and press $\boxed{ENTER}$. "Plot 1" should be highlighted. Before viewing the plot any existing entries on the equation-editor screen should be cleared. The easiest way to choose a viewing window is to use the ZOOMSTAT feature. Press $\boxed{ZOOM}$ $\boxed{9}$ or $\boxed{ZOOM}$ and the down-arrow until the selection cursor is beside "9:ZoomStat" and press $\boxed{ENTER}$. The ZOOMSTAT feature redefines the viewing window to display all statistical data points.

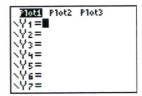  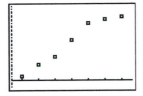

With all the data points in the viewing window it is easy to press $\boxed{WINDOW}$ and modify the window to suit your needs. Below the window has been changed and the new graph is displayed.

 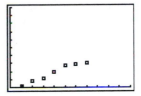

From the scatter plot, the data appears to be approximately linear.

(b)   Find a linear function that (approximately) fits the data.

The calculator's LINEAR REGRESSION feature can be used to fit a linear equation to the data. From the home screen and with the data points in the lists, press $\boxed{STAT}$ $\boxed{\blacktriangleright}$ $\boxed{4}$ to select "4:LinReg($ax + b$)" from the STATISTICAL CALCULATIONS menu. Now press $\boxed{2nd}$ $\boxed{L1}$ $\boxed{,}$ $\boxed{2nd}$ $\boxed{L2}$ to let the calculator know which lists contain the data. (Even though the calculator was programmed to assume the x- and y-coordinates of data points are in L_1 and L_2, respectively, it is good practice to specify the lists being used.) Before the regression equation is found, it is possible to select a y-variable to which it will be stored on the equation-editor screen. To do this, press $\boxed{,}$ $\boxed{VARS}$ $\boxed{\blacktriangleright}$ $\boxed{ENTER}$ $\boxed{ENTER}$. Finally, to display the coefficients of the regression equation press $\boxed{ENTER}$ again.

If the diagnostics have been turned on in your calculator, values of r^2 and r will also be displayed. These numbers indicate how well the regression line fits the data. For the remainder of this manual, regression will be done with the diagnostics turned off.

  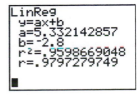 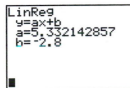

Rounding the coefficient of x to the nearest hundredth, we have $y = 5.33x - 2.8$.

If you wish to select DIAGNOSTIC ON mode, press [2nd] [CATALOG] and use [▼] to position the triangular selection cursor beside "DiagnosticOn." To alleviate the tedium of scrolling through many items to reach "DiagnosticOn," press D after pressing [2nd] [CATALOG] to move quickly to the first catalog item that begins with the letter D. (D is the alphabetic character associated with the [x⁻¹] key.) Then use [▼] to scroll to "DiagnosticOn." Note that it is not necessary to press [ALPHA] D. Press [ENTER] to paste this instruction on the home screen and then press [ENTER] a second time to set the mode. To select DIAGNOSTIC OFF mode, press [2nd] [CATALOG], position the selection cursor beside "DiagnosticOff," press [ENTER] to paste the instruction on the home screen, and then press [ENTER] again to set the mode.

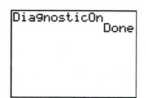   

Press [GRAPH] to see the line displayed with the data points.

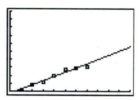

(c)   Use the model to predict the percentage by which 2010 wages will exceed 1996 wages.

To predict the percentage by which 2010 wages will exceed 1996 wages, evaluate the regression equation for $x = 14$. (2010 is 14 years after 1996.) Use any of the methods for evaluating a function presented on page 9 of this manual. We will use function notation on the home screen. Press [VARS] [▶] [ENTER] [ENTER] [(] [1] [4] [)] [ENTER].

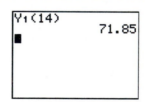

When $x = 14$, $y \approx 71.85$, so we predict 2010 wages will exceed 1996 wages by approximately 71.85%.

POLYNOMIAL REGRESSION; Quadratic and Cubic

Both the TI-83 Plus and TI-84 Plus graphing calculators have the capability to use the REGRESSION feature to fit quadratic, cubic, and other functions to data.

Section R.6, Technology Connection, page 81   The following chart relates the number of live births to women of a particular age.

Age, x	Average Number of Live Births Per 1000 Women
16	34
18.5	86.5
22	111.1
27	113.9
32	84.5
37	35.4
42	6.8

(a)　　　Fit a quadratic function to the data using the REGRESSION feature on the calculator. Then make a scatter plot of the data and graph the quadratic function with the scatter plot.

Clear any existing entries on the equation-editor screen. Enter the data into lists with the ages in L_1 and the average number of live births per 1000 women in L_2. To do this, press [STAT] [ENTER] to get into the statistical list-editor screen. Clear any data in L_1 and L_2 by using the arrow keys to position the cursor over the name of the list to be cleared. Then press [CLEAR] [ENTER]. Once both lists are cleared enter the data as we have done here. Then press [STAT] [▶] [5] to select the QUADRATIC REGRESSION feature, denoted "5:QuadReg," from the STATISTICAL CALCULATIONS menu. Now to tell the calculator which lists contain the data, press [2nd] [L1] [,] [2nd] [L2]. (L_1 and L_2 are the factory defaults.) To see the graph, press [,] [VARS] [▶] [ENTER] [ENTER]. This will put the quadratic regression, not a rounded off version of the function, on the equation-editor screen to the right of "$Y_1 =$" so it can be graphed. Press [ENTER] and the calculator returns the coefficients of a quadratic function of the form $f(x) = ax^2 + bx + c$. Rounding the coefficients to two decimal places, we obtain the function $f(x) = -0.49x^2 + 25.95x - 238.49$.

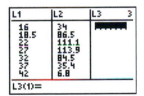  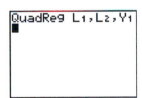 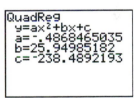

We also want to see the data plotted. Press [2nd] [STAT PLOT] [ENTER] [ENTER] to turn on the statplot. To have the calculator set the window press [ZOOM] [9].

 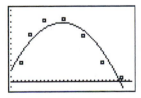

(b)　　　Fit a cubic function to the data using the REGRESSION feature on the calculator. Then make a scatter plot of the data and graph the cubic function with the scatter plot.

Once the data are entered, fit a cubic function to it by pressing [STAT] [▶] [6] [2nd] [L1] [,] [2nd] [L2] [,] [VARS] [▶] [ENTER] [2]. These keystrokes select the CUBIC REGRESSION feature, denoted "6:CubicReg" from the

STATISTICAL CALCULATIONS menu, tell the calculator which lists contain the data, and tell the calculator to put the equation to the right of "$Y_2 =$" on the equation-editor screen. Press $\boxed{\text{ENTER}}$ and the calculator displays the coefficients of a cubic function of the form $f(x) = ax^3 + bx^2 + cx + d$. Rounding the coefficients to two decimal places we obtain the function $f(x) = 0.03x^3 - 3.22x^2 + 101.18x - 886.93$.

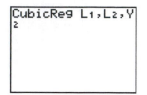   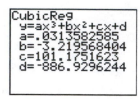

Before graphing the cubic function we need to turn off the quadratic function. Press $\boxed{\text{Y=}}$ to see both the quadratic and cubic equations on the equation-editor screen. Position the cursor over "=" in "$Y_1 =$" and press $\boxed{\text{ENTER}}$. This deselects the function, or turns it off. When we graph, it will not be shown. Since Plot1 is already turned on (notice Plot 1 is highlighted), and we already have a suitable window, press $\boxed{\text{GRAPH}}$ to see the data points and the cubic function graphed together.

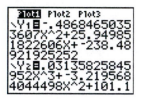   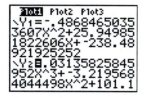   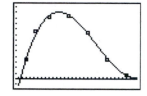

(c)   Which function seems to fit the data better?

The graph of the cubic function appears to fit the data better than the graph of the quadratic function. Remember that a prediction is only an approximation.

(d)   Use the function from part (c) to estimate the average number of live births by women of ages 20 and 30.

We can use any of the methods described on page 9 of this manual to evaluate the function. We will use function notation on the home screen. Press $\boxed{\text{VARS}}$ $\boxed{\blacktriangleright}$ $\boxed{\text{ENTER}}$ $\boxed{2}$ $\boxed{(}$ $\boxed{2}$ $\boxed{0}$ $\boxed{)}$ $\boxed{\text{ENTER}}$. We estimate that there will be about 100 live births per 1000 20-year-old women. Press $\boxed{\text{VARS}}$ $\boxed{\blacktriangleright}$ $\boxed{\text{ENTER}}$ $\boxed{2}$ $\boxed{(}$ $\boxed{3}$ $\boxed{0}$ $\boxed{)}$ $\boxed{\text{ENTER}}$. We estimate that there will be about 97 live births per 1000 30-year-old women.

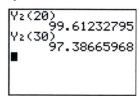

Chapter 1
Differentiation

ZOOM IN FEATURE

The ZOOM IN feature from the ZOOM menu can be used to enlarge a portion of a graph.

Section 1.2, Technology Connection, page 123 Verify graphically that $\lim\limits_{x \to 0} \dfrac{\sqrt{x+1}-1}{x} = 0.5$.

First, graph $Y_1 = \dfrac{\sqrt{x+1}-1}{x}$ in a window that shows the portion of the graph near $x = 0$. One good choice

for the window is [-2, 5, -1, 2]. This function requires several sets of parenthesis. Look at the equation-editor

screen below to make sure you have used the parenthesis correctly. Press $\boxed{\text{TRACE}}$ and use the $\boxed{\blacktriangleleft}$ and $\boxed{\blacktriangleright}$ keys to move

the cursor to a point on the curve near $x = 0$.

  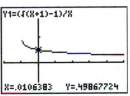

Now select the ZOOM IN feature from the ZOOM menu by pressing $\boxed{\text{ZOOM}}$ $\boxed{2}$ $\boxed{\text{ENTER}}$. This enlarges the

portion of the graph near $x = 0$. We can now press $\boxed{\text{TRACE}}$ again and use the left- and right-arrow keys to trace the

curve near $x = 0$.

 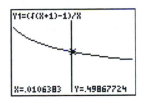 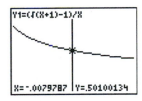

The ZOOM IN feature can be used as many times as desired in order to verify that the limit as x approaches

0 is .5.

THE TABLE FEATURE; G-T MODE

For an equation entered in the equation-editor screen, a TABLE of x- and y-values and the graph can be

displayed side by side using GRAPH-TABLE mode.

Section 1.2, Technology Connection, page 123 Verify graphically that $\lim\limits_{x \to 0} \dfrac{\sqrt{x+1}-1}{x} = 0.5$.

First, graph $Y_1 = \dfrac{\sqrt{x+1}-1}{x}$ in a window that shows the portion of the graph near $x = 0$. One good choice

for the window is [-2, 5, -1, 2]. Using the GRAPH-TABLE mode, we can view both the graph and the table at the same time. Press $\boxed{\text{MODE}}$, move the cursor over "G-T" and press $\boxed{\text{ENTER}}$. In this instance the screens look a little different depending on the calculator you are using.

Press $\boxed{\text{2nd}}$ $\boxed{\text{TBLSET}}$. Use the settings on the screen below on the left. The TABLE feature is explained on page 8 of this manual. Press $\boxed{\text{GRAPH}}$ and see the screen below on the right.

   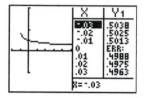

This is a way to verify that the limit as x approaches 0 is .5.

THE nDeriv FEATURE

The NUMERICAL DERIVATIVE feature, "nDeriv(," from the MATH menu can be used to evaluate a numerical derivative on the home screen. Remember that the numerical derivative is the slope of the tangent line to a curve at a specific point.

Section 1.4, Technology Connection, page 144   For the function $f(x) = x(100 - x)$, find $f'(70)$.

The NUMERICAL DERIVATIVE feature is item 8 on the MATH submenu of the MATH menu. Press $\boxed{\text{MATH}}$ $\boxed{8}$ or press $\boxed{\text{MATH}}$, highlight item 8, and press $\boxed{\text{ENTER}}$ to select "8:nDeriv(." Now enter the function by pressing $\boxed{\text{X,T,}\theta,n}$ $\boxed{(}$ $\boxed{1}$ $\boxed{0}$ $\boxed{0}$ $\boxed{-}$ $\boxed{\text{X,T,}\theta,n}$ $\boxed{)}$ $\boxed{,}$, the variable by pressing $\boxed{\text{X,T,}\theta,n}$ $\boxed{,}$, and the value at which the derivative is to be evaluated by pressing $\boxed{7}$ $\boxed{0}$ $\boxed{)}$. Notice that the fields are separated by commas. Press $\boxed{\text{ENTER}}$ and the numerical derivative is given. Note that the graphing calculator supplies a left parenthesis after "nDeriv" and we supply the right parenthesis after entering 70.

   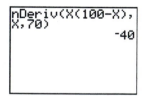

If the function is first entered into Y_1 on the equation-editor screen as in the screen on the next page on the left, we can evaluate the numerical derivative by going to the home screen and pressing $\boxed{\text{MATH}}$ $\boxed{8}$ $\boxed{\text{VARS}}$ $\boxed{\blacktriangleright}$ $\boxed{\text{ENTER}}$ $\boxed{\text{ENTER}}$ $\boxed{,}$ $\boxed{\text{X,T,}\theta,n}$ $\boxed{,}$ $\boxed{7}$ $\boxed{0}$ $\boxed{)}$ $\boxed{\text{ENTER}}$.

   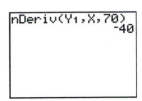

We see that $f'(70) = -40$. Since this is the value of the numerical derivative at $x = 70$, we know that the slope of the tangent line to the curve at $x = 70$ is -40.

THE TANGENT FEATURE; On the Home Screen

We can draw a line tangent to a curve at a given point using the TANGENT feature from the DRAW submenu of the DRAW menu.

Section 1.4, Technology Connection, page 144   Draw the line tangent to the graph of $f(x) = x(100-x)$ at $x = 70$.

First, graph $Y_1 = x(100-x)$ in a window that shows the portion of the curve near $x = 70$. One good window choice is shown below. Be sure to clear any functions that were previously entered and turn off the plots.

      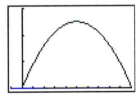

Now press [2nd] [QUIT] to go to the home screen. Here we will select "5:Tangent(" from the DRAW menu and instruct the calculator to draw the line tangent to the graph of Y_1 at $x = 70$. To do, this press [2nd] [DRAW] [5] [VARS] [▶] [ENTER] [ENTER] [,] [7] [0] [)] [ENTER]. ([DRAW] is the second function associated with the [PRGM] key.) Note that the calculator supplies a left parenthesis after "Tangent" and we close the parentheses with a right parenthesis after entering 70.

   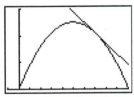

The keystrokes [2nd] [DRAW] [5] access the DRAW menu and choose "5:Tangent(." The same result can be achieved by pressing [2nd] [DRAW], using the down-arrow to highlight item 5, and then pressing [ENTER]. To clear the drawing from the graph screen, use the "1:ClrDraw" (clear drawing) command from the DRAW menu. Press [2nd] [DRAW] [ENTER] or [2nd] [DRAW] [1].

See page 33 of this manual for a procedure that draws a tangent line and produces the equation of the tangent line directly from the graph screen.

THE $\frac{dy}{dx}$ FEATURE

The NUMERICAL DERIVATIVE feature, "dy/dx," from the CALCULATE menu can be used to evaluate a numerical derivative on the graph screen. Remember that the numerical derivative is the slope of the tangent line to a curve at a specific point. We can find the derivative of a function at a specific point directly from the graph screen.

Section 1.5, Technology Connection, page 149 For the function $f(x) = x\sqrt{4 - x^2}$, find $\frac{dy}{dx}$ at a specific point.

First graph $Y_1 = x\sqrt{4 - x^2}$. We will use the window [-3, 3, -4, 4]. Then select "6: dy/dx" from the CALCULATE menu by pressing [2nd] [CALC] [6] or by pressing [2nd] [CALC] to access the CALCULATE menu, using the [▼] key to highlight "6: dy/dx," and then pressing [ENTER]. We see the graph with a cursor positioned on the x-value that is the x-coordinate of the midpoint of the x-axis.

 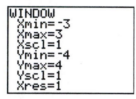  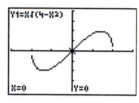

To find the value of dy/dx at a specific point either move the cursor to the desired point or key in the point's x-coordinate. For example, press [2nd] [CALC] [6], as explained above, and move the cursor to the point (1.212766, 1.9287169). Now press [ENTER] and the derivative is displayed. We see that $dy/dx = 0.66551339$ at this point.

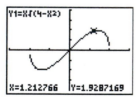 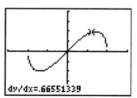

To evaluate dy/dx for $x = 1$, select "dy/dx" from the CALCULATE menu by pressing [2nd] [CALC] [6] and then press [1] [ENTER].

 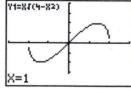 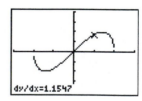

We see that $dy/dx = 1.1547$, when $x = 1$.

THE TANGENT FEATURE; On the Graph Screen

Section 1.5, Technology Connection, page 149 Draw the line tangent to the graph of $f(x) = x\sqrt{4-x^2}$ at a specific point.

First, graph $f(x) = x\sqrt{4-x^2}$ in an appropriate viewing window. Here we are using [-3, 3, -4, 4]. Now select "5:Tangent(" from the DRAW submenu of the DRAW menu by pressing [2nd] [DRAW] [5] use the left- and right-arrow keys to move the cursor to the desired point. Press [ENTER] and see the tangent line drawn at that point. The first coordinate of the point of tangency and the equation of the tangent line are also displayed.

 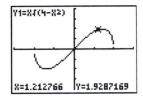 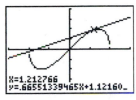

Use "1:ClrDraw" (clear drawing) from the DRAW menu to clear the graph of the tangent line from the graph screen. This can be done two ways. If the graph and the tangent line are still displayed, press [2nd] [DRAW] [ENTER] or [2nd] [DRAW] [1] and the graph will be redrawn without the tangent line. If the graph and the tangent line are still on the graph screen, but the home screen is displayed, clear the drawing by pressing [2nd] [DRAW] [ENTER] [ENTER] or [2nd] [DRAW] [1] [ENTER]. Either of these methods will result in the graph without the tangent line being displayed on the graph screen.

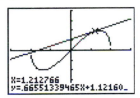 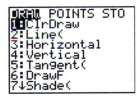 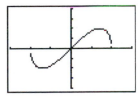

We can enter the x-coordinate of the point where we want the tangent line to be drawn. For example, to graph the line tangent to the curve at $x = 1$, from the graph screen press [2nd] [DRAW] [5] as before to select "5:Tangent(." Press [1] [ENTER] to enter 1 as the x-coordinate.

 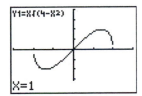 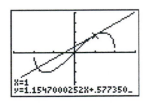

Use "1:ClrDraw" (clear drawing) from the DRAW menu to clear the graph of the tangent line from the graph screen. This can be done two ways. If the graph and the tangent line are still displayed, press [2nd] [DRAW] [ENTER] or [2nd] [DRAW] [1] and the graph will be redrawn without the tangent line. If the graph and the tangent line are still on the graph screen, but the home screen is displayed, clear the drawing by pressing [2nd] [DRAW] [ENTER] [ENTER] or

2nd [DRAW] 1 [ENTER]. Either of these methods will result in the graph without the tangent line being displayed on the graph screen.

If the arrow keys have been used to position the cursor at a point when using the "dy/dx" operation from the CALCULATE menu, the tangent line at that point can be graphed from the graph screen immediately after the value of dy/dx is displayed by pressing 2nd [DRAW] 5 [ENTER].

ENTERING THE SUM OF TWO FUNCTIONS

The Y-VARIABLES submenu of the VARIABLES menu can be used to enter a sum of functions.

Section 1.5, Technology Connection, page 152 Given that $Y_1 = x(100 - x)$, $Y_2 = x\sqrt{100 - x^2}$, and $Y_3 = Y_1 + Y_2$, find the derivative of each of the three functions at $x = 8$.

First press Y= to go to the equation-editor screen. Clear any entries that are present and turn off the plots. Then enter $Y_1 = x(100 - x)$ and $Y_2 = x\sqrt{100 - x^2}$. To enter $Y_3 = Y_1 + Y_2$, we select the functions Y_1 and Y_2 from the Y-VARIABLES submenu of the VARIABLES menu. Position the cursor to the right of "Y_3=" and press VARS ▶ 1 1 + VARS ▶ 1 2.

   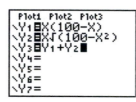

Press 2nd [QUIT] to go to the home screen. We will use the NUMERICAL DERIVATIVE function to find the derivative of each of the three functions at $x = 8$. Select this operation by pressing MATH 8 or by pressing MATH and using the down-arrow to highlight item 8 and then pressing [ENTER]. These keystrokes copy "nDeriv(" to the home screen. Now enter Y_1 by pressing VARS ▶ 1 1 , , the variable by pressing X,T,Θ,n , , and the value at which the derivative is to be evaluated by pressing 8). Notice that the fields are separated by commas. Press [ENTER] and the numerical derivative is given. Note that the graphing calculator supplies a left parenthesis after "nDeriv" and we supply the right parenthesis after entering 8.

   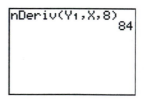

On the home screen do the next numerical derivative by pressing MATH 8. Now enter Y_2 by pressing VARS ▶ 1 2 , , the variable by pressing X,T,Θ,n , , and the value at which the derivative is to be evaluated by pressing 8). Press [ENTER] and the numerical derivative is given.

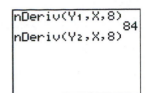 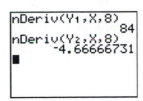

On the home screen do the next numerical derivative by pressing [MATH] [8]. Now enter Y_3 by pressing [VARS] [▶] [1] [3] [,], the variable by pressing [X,T,Θ,n] [,], and the value at which the derivative is to be evaluated by pressing [8] [)]. Press [ENTER] and the numerical derivative is given.

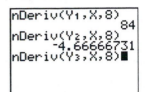 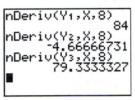

Notice that the third derivative is the sum of the first two.

DESELECTING FUNCTIONS; GRAPH STYLES

Several functions can be graphed at the same time and they can be graphed using different graph styles. Functions can be deselected on the equation-editor so that they remain there, but will not be graphed.

Section 1.6, Technology Connection and Example 5, page 162 We have the original function, $Y_1 = \dfrac{x^2 - 3x}{x - 1}$, the derivative found in Example 5 on page 162 in your text, $Y_2 = \dfrac{x^2 - 2x + 3}{(x-1)^2}$, and the derivative the calculator will graph, $Y_3 = \text{nDeriv}(Y_1, x, x)$. If the graphs of Y2 and Y_3 coincide, the answer in Example 5 is verified by the calculator. First enter $Y_1 = \dfrac{x^2 - 3x}{x - 1}$ and $Y_2 = \dfrac{x^2 - 2x + 3}{(x-1)^2}$. Now enter $Y_3 = \text{nDeriv}(Y_1, x, x)$. Position the cursor immediately to the right of "Y_3=" and press [MATH] [8] [VARS] [▶] [1] [1] [,] [X,T,Θ,n] [,] [X,T,Θ,n] [)].

Since we want to see only the graphs of Y_2 and Y_3, we will deselect Y_1. Position the cursor over "=" in "Y_1=" and press [ENTER]. This deselects the function, or turns it off so it will not be graphed.

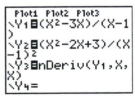

To select Y_1 again, position the cursor over "=" in "Y_1=" and press ⎣ENTER⎦. The equals sign will be highlighted now, indicating that the function is selected. For now, leave it deselected.

With Y_1 deselected, we can graph Y_2 and Y_3 using different graph styles to determine whether the graphs coincide. When the graphing calculator is in CONNECTED mode, equations are graphed with a solid line. You can see the graph style icon to the left of each of the y-variable names looks like a solid line.

   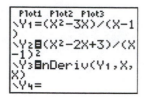

Make sure the calculator is in SEQUENTIAL mode before proceeding. Press ⎣MODE⎦ and make sure "SEQUENTIAL" is highlighted. We will keep the solid line graph style for the graph of Y_2 and select the path style for the graph of Y_3. After the graph of Y_2 is drawn, a circular cursor will trace over the graph that is already displayed. To select the path style for Y_3, position the cursor over the graph style icon to the left of "Y_3=" and press ⎣ENTER⎦ four times until you see "⊸."

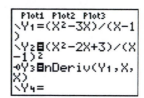

Press ⎣ZOOM⎦ ⎣6⎦. The cursor that resembles a ball moves along the first graph. We could also examine the table to verify the results. Below are the table settings and the table these settings produce.

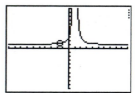   

Both verify that the derivative found in Example 5 is correct.

Chapter 2
Applications of Differentiation

THE MAXIMUM and MINIMUM FEATURES

Section 2.1, Technology Connection, page 210 Use the MAXIMUM and MINIMUM features from the CALCULATE menu to approximate the relative extrema of $f(x) = -0.4x^3 + 6.2x^2 - 11.3x - 54.8$ on the graph screen.

First, graph $Y_1 = -0.4x^3 + 6.2x^2 - 11.3x - 54.8$ in a window that displays the relative extrema of the function. The window settings shown below were found using trial and error. Observe that a relative maximum occurs near x = 10 and a relative minimum occurs near x = 1.

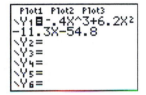 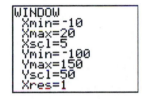 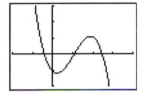

To find the relative maximum, first press [2nd] [CALC] [4] or [2nd] [CALC] [▼] [▼] [▼] [ENTER] to select "4:maximum" from the CALCULATE menu. We are prompted to select a left bound for the relative maximum. This means that we must choose an *x*-value that is to the left of the *x*-coordinate of the point where the relative maximum occurs. This can be done by using the left- and right-arrow keys to move the cursor to a point to the left of the relative maximum or by keying in an appropriate value.

 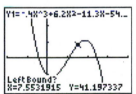

Once this is done, press [ENTER]. Notice that a triangular marker has appeared to mark the left bound. Now we are prompted to select a right bound. Either move the cursor to a point to the right of the relative maximum or key in an appropriate value.

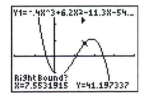 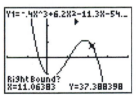

Press [ENTER]. Notice another triangular marker has appeared to mark the right bound. These two markers provide boundaries for the calculator. Finally we are prompted to guess the *x*-value at which the relative maximum occurs. Move the cursor near the relative maximum point or key in an *x*-value. Press [ENTER] a third time. We see that the relative maximum function value is approximately 54.61 when $x \approx 9.32$.

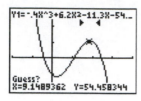   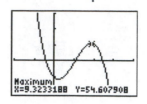

To find the relative minimum, select "3:minimum" from the CALCULATE menu by pressing [2nd] [CALC] [3] or [2nd] [CALC] [▼] [▼] [ENTER]. Select left and right bounds for the relative minimum and guess the *x*-value at which it occurs as described above.

 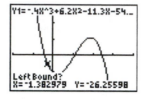 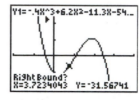 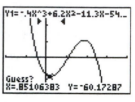

We see that a relative minimum function value of approximately -60.30 occurs when $x \approx 1.01$.

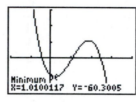

To summarize, a relative maximum function value is approximately 54.61 when $x \approx 9.32$ and a relative minimum function value of approximately -60.30 occurs when $x \approx 1.01$.

THE fMax AND fMin FEATURES

The FUNCTION MAXIMUM and FUNCTION MINIMUM features from the MATH menu can be used on the home screen to calculate the *x*-values at which relative maximum and minimum values of a function occur over a specified closed interval.

Section 2.1, Technology Connection, page 210 Use the FUNCTION MAXIMUM and FUNCTION MINIMUM features to approximate the relative extrema of $f(x) = -0.4x^3 + 6.2x^2 - 11.3x - 54.8$ on the home screen.

First enter the function as Y_1 and graph it as shown below. Observe that a relative maximum occurs in the closed interval [5, 15]. There are other intervals we could use. Keep in mind that the larger the interval, the longer it takes the calculator to return an *x*-value.

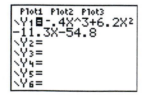 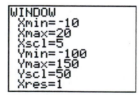 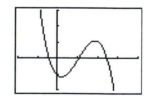

Now press [2nd] [QUIT] to go to the home screen and then press [MATH] [7] to select "7:fMax(" from the MATH submenu of the MATH menu. Enter the name of the function, the variable, and the left and right endpoints of the interval on which the relative maximum occurs by pressing [VARS] [▶] [1] [1] [,] [X,T,Θ,n] [,] [5] [,] [1] [5] [)]. Press

[ENTER] to find that the relative maximum occurs when $x \approx 9.323320694$. To find the relative maximum value of the function, we evaluate the function for this value of x. Press [VARS] [▶] [1] [1] [(] [2nd] [ANS] [)] [ENTER]. ([ANS] is the second function associated with the [(-)] key.) The keystrokes [2nd] [ANS] cause the calculator to use the previous answer, 9.323320694, as the value for x in Y_1.

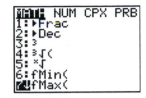 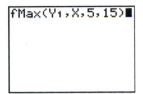 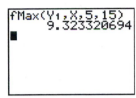 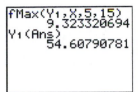

We also observe that a relative minimum occurs in the interval [-5, 5]. Again, there are other intervals we could choose. To find the relative minimum in this interval, first press [MATH] [6] to select "6:fMin(" from the MATH submenu of the MATH menu. Then enter the name of the function, the variable, and the endpoints of the interval. Press [VARS] [▶] [1] [1] [,] [X,T,Θ,n] [,] [(-)] [5] [,] [5] [)]. Press [ENTER] to find that the relative minimum occurs when $x \approx 1.010010343$. To find the relative minimum value of the function, we evaluate the function for this value of x. Press [VARS] [▶] [1] [1] [(] [2nd] [ANS] [)] [ENTER]. ([ANS] is the second function associated with the [(-)] key.) The keystrokes [2nd] [ANS] cause the calculator to use the previous answer, 1.010010343, as the value for x in Y_1.

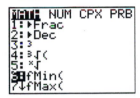 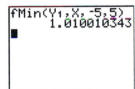 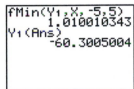

To summarize, a relative maximum function value is approximately 54.61 when $x \approx 9.32$ and a relative minimum function value of approximately -60.30 occurs when $x \approx 1.01$.

LIMITS AT INFINITY and HORIZONTAL ASYMPTOTES

Section 2.3, Technology Connection, page 234 Use function notation to verify that $\lim\limits_{x \to \infty} \dfrac{3x-4}{x} = 3$ and find the function's horizontal asymptote.

With $Y_1 = \dfrac{3x-4}{x}$, go to the home screen and press [VARS] [▶] [ENTER] [ENTER] [(] here we want to put a really large number (approaching infinity). Press [1] [0] continue pressing [0] 15 or 16 times. Then press [)] [ENTER].

 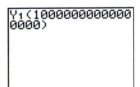 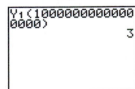

We see that the limit at infinity is 3. Also, the horizontal asymptote is the line $y = 3$.

THE MAXIMUM and MINIMUM FEATURES

Section 2.4, Technology Connection and Example 1, page 248   Use the MAXIMUM and MINIMUM features

from the CALCULATE menu to approximate the absolute extrema of $f(x) = x^3 - 3x + 2$ over the interval $\left[-2, \dfrac{3}{2}\right]$

on the graph screen.

First graph $Y_1 = x^3 - 3x + 2$ in a window with $\text{Xmin} = -2$ and $\text{Xmax} = \dfrac{3}{2}$. Over this interval an absolute

maximum occurs near $x = -1$ and an absolute minimum occurs near $x = 1$ and $x = -2$. Since we have a window

which displays the entire interval we are interested in, it is obvious that the endpoint $x = \dfrac{3}{2}$ is not the x-coordinate of

an absolute maximum or minimum point over this interval.

 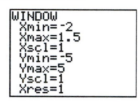 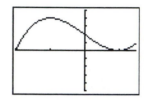

To find the absolute maximum, press [2nd] [CALC] [4] or [2nd] [CALC] [▼] [▼] [▼] [ENTER] to select "4:maximum"

from the CALCULATE menu. We are prompted to select a left bound for the absolute maximum. This means that

we must choose an x-value that is to the left of the x-coordinate of the point where the absolute maximum occurs.

This can be done by using the left- and right-arrow keys to move the cursor to a point to the left of the absolute

maximum or by keying in an appropriate value.

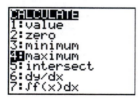 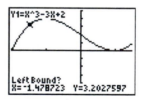

Once this is done, press [ENTER]. Notice that a triangular marker has appeared to mark the left bound. Now

we are prompted to select a right bound. We move the cursor to a point to the right of the absolute maximum or we

key in an appropriate value.

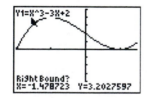 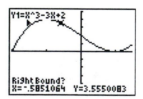

Press [ENTER]. Notice another triangular marker has appeared to mark the right bound. These two markers

provide boundaries for the calculator. Finally we are prompted to guess the x-value at which the absolute maximum

occurs. Move the cursor near the absolute maximum point or key in an x-value. Press [ENTER] a third time. We see

that the absolute maximum function value is 4 when $x = -1$.

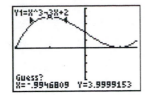 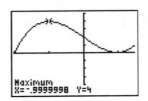

To find the absolute minimum, select "3:minimum" from the CALCULATE menu by pressing [2nd] [CALC] [3] or [2nd] [CALC] [▼] [▼] [ENTER]. Select left and right bounds for the absolute minimum and guess the *x*-value at which it occurs as described above.

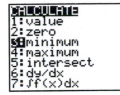 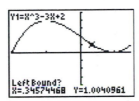 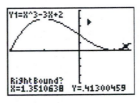 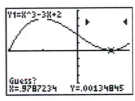

On this interval, we have an absolute minimum function value of 0 when $x = 1$.

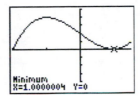

We are also obligated to check the endpoints of the interval $\left[-2, \dfrac{3}{2} \right]$. To do this we use the VALUE

feature from the CALCULATE menu. With the graph of the function displayed, press [2nd] [CALC] [ENTER] or [2nd] [CALC] [1] to choose "1:Value" from the CALCULATE menu. We must enter the *x*-value, -2, and then press [ENTER]. There is a second absolute minimum function value of 0. It occurs when $x = -2$.

 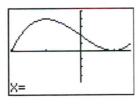 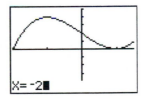 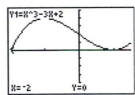

We can see that the other endpoint is neither a maximum nor a minimum, but we repeat the procedure described above for completeness. Note that if we have just used the VALUE feature, we only have to enter the next *x*-value we want to evaluate; we do not have to return to the CALCULATE menu.

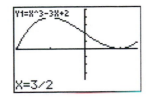 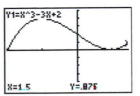

To summarize, the absolute maximum value of the function over the interval is 4 when $x = -1$. The absolute minimum value of the function over the interval is 0 when $x = 1$ or when $x = -2$.

THE fMax and fMin FEATURES

The FUNCTION MAXIMUM and FUNCTION MINIMUM features from the MATH menu can be used on the home screen to calculate the x-values at which absolute maximum and minimum values of a function occur over a specified closed interval.

Section 2.4, Technology Connection and Example 1, page 248   Use the FUNCTION MAXIMUM and FUNCTION MINIMUM features of a graphing calculator to approximate the absolute extrema of

$$f(x) = x^3 - 3x + 2 \text{ over the interval} \left[-2, \frac{3}{2}\right] \text{ on the home screen.}$$

First enter the function as Y_1 and graph it as shown below.

 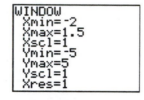 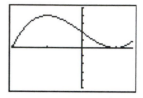

Press [2nd] [QUIT] to go to the home screen and then press [MATH] [7] to select "7:fMax(" from the MATH submenu of the MATH menu. Enter the name of the function, the variable, and the left and right endpoints of the interval on which the absolute maximum occurs by pressing [VARS] [▶] [1] [1] [,] [X,T,Θ,n] [,] [(-)] [2] [,] [3] [÷] [2] [)]. Press [ENTER] to find that the absolute maximum occurs when $x = -1$. To find the absolute maximum value of the function, we evaluate the function for this value of x. Press [VARS] [▶] [1] [1] [(] [2nd] [ANS] [)] [ENTER]. ([ANS] is the second function associated with the [(-)] key.) The keystrokes [2nd] [ANS] cause the calculator to use the previous answer, -1, as the value for x in Y_1.

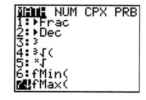 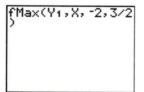 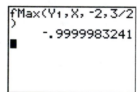 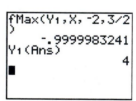

We also observe that two absolute minimums occur in the interval $\left[-2, \frac{3}{2}\right]$. We will have to use different intervals in order to find both of them. The calculator will find one of them if we use the entire interval, so press [MATH] [6] to select "6:fMin(" from the MATH submenu of the MATH menu. Then enter the name of the function, the variable, and the endpoints of the interval. Press [VARS] [▶] [1] [1] [,] [X,T,Θ,n] [,] [(-)] [2] [,] [3] [÷] [2] [)]. Press [ENTER]. One of the absolute minimums occurs when $x = 1$. To find the absolute minimum value of the function, we evaluate the function for this value of x. Press [VARS] [▶] [1] [1] [(] [2nd] [ANS] [)] [ENTER]. ([ANS] is the second function associated with the [(-)] key.) The keystrokes [2nd] [ANS] cause the calculator to use the previous answer, 1, as the value for x in Y_1.

 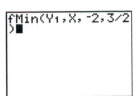 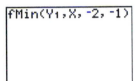 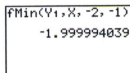

To find the second absolute minimum over the given interval, we must change the interval previously used in the FUNCTION MINIMUM, "fMin(," feature. The new interval must not contain the x-value we just found, that is to say, 1 can not be in the new interval. Press MATH 6 to select "6:fMin(" from the MATH submenu of the MATH menu. Then enter the name of the function, the variable, and the endpoints of the interval. We are using the closed interval from -2 to -1. Press VARS ▶ 1 1 , X,T,Θ,n , (-) 2 , (-) 1). Press ENTER. The second absolute minimum occurs when $x = -2$. To find the absolute minimum value of the function, we evaluate the function for this value of x. Press VARS ▶ 1 1 (2nd [ANS]) ENTER. ([ANS] is the second function associated with the (-) key.) The keystrokes 2nd [ANS] cause the calculator to use the previous answer, -2, as the value for x in Y_1.

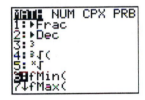 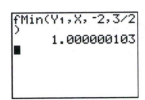

NOTE: The number 5.3647537E-5 is scientific notation on the calculator. It means 5.3647537×10^{-5}. This number written as a decimal is 0.000053647537. This is very nearly 0.

To summarize, the absolute maximum value of the function over the interval is 4 when $x = -1$. The absolute minimum value of the function over the interval is 0 when $x = 1$ or $x = -2$.

THE LISTS (as a spread sheet)

The lists can be used like a spread sheet to solve problems that involve mathematical formulas.

Section 2.5, Technology Connection, page 260

1.   Use the lists to complete the table: Using the lists in the calculator, this table becomes:

x	$y = 20 - x$	$A = x(20 - x)$
0		
4		
6.5		
8		
10		
12		
13.2		
20		

x L_1	$y = 20 - x$ $L_2 = 20 - L_1$	$A = x(20 - x)$ $L_3 = L_1 * L_2$
0		
4		
6.5		
8		
10		
12		
13.2		
20		

To clear existing entries in the lists press [STAT] [4] [2nd] [L1] [,] [2nd] [L2] [,] [2nd] [L3] [,] [2nd] [L4] [,] [2nd] [L5] [,] [2nd] [L6] [ENTER]. (L$_1$ through L$_6$ are the second functions associated with the numeric keys [1] through [6].) The lists can also be cleared by first accessing the statistical list-editor screen by pressing [STAT] [ENTER] or [STAT] [1] to select "1:Edit" from the menu. Then, for each list that contains entries, use the arrow keys to move the cursor to highlight the name of the list at the top of the column and press [CLEAR] [▼] or [CLEAR] [ENTER].

Once the lists are cleared, we can enter the x-values in L$_1$. After entering each value, press [ENTER] or [▼] to drop down to the next line and make the next entry in the list. This list is too long to be seen on a single screen. We can use the [▲] and [▼] keys to scroll up and down in the list and view all the data.

With the data in, move the cursor to highlight "L$_2$" and press [2] [0] [–] [2nd] [L1]. ([L1] is the second function associated with the [1] key.) Press [ENTER] and watch L$_2$ fill in.

Now we move the cursor to highlight "L$_3$" and press [2nd] [L1] [×] [2nd] [L2]. ([L2] is the second function associated with the [2] key.) Press [ENTER] and watch L$_3$ fill in.

We can use these to fill in the original table. Here is the table filled in completely.

x L$_1$	$y = 20 - x$ L$_2$ $= 20 - $ L$_1$	$A = x(20 - x)$ L$_3$ = L$_1$ * L$_2$
0	20	0
4	16	64
6.5	13.5	87.75
8	12	96
10	10	100
12	8	96
13.2	6.8	89.76
20	0	0

From this table, we see that the maximum area is 100 square feet. This maximum area occurs when both the length and width are 10 feet.

2.   Graph $A(x) = x(20 - x)$ over the interval [0, 20].

Press $\boxed{Y=}$ and clear existing entries and turn off the plots. The equation becomes $Y_1 = x(20 - x)$. The y-value in the calculator is the area and the x-value is the x-value in the equation. Below we have the equation-editor screen and the window screen. Press $\boxed{GRAPH}$ to see the area function graphed.

   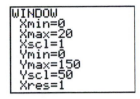   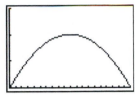

3. Estimate a maximum value of the function and tell the x-value that generates it.

To find the maximum, first press $\boxed{2nd}$ $[CALC]$ $\boxed{4}$ or $\boxed{2nd}$ $[CALC]$ $\boxed{\blacktriangledown}$ $\boxed{\blacktriangledown}$ $\boxed{\blacktriangledown}$ $\boxed{ENTER}$ to select "4:maximum" from the CALCULATE menu. We are prompted to select a left bound for the maximum. This means that we must choose an x-value that is to the left of the x-coordinate of the point where the absolute maximum occurs. This can be done by using the $\boxed{\blacktriangleleft}$ and $\boxed{\blacktriangleright}$ keys to move the cursor to a point to the left of the absolute maximum or by keying in an appropriate value.

   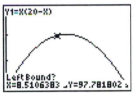

Once this is done, press $\boxed{ENTER}$. Notice that a triangular marker has appeared to mark the left bound. Now we are prompted to select a right bound. Move the cursor to a point to the right of the absolute maximum or key in an appropriate value.

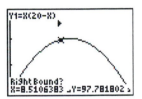   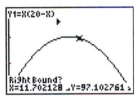

Press $\boxed{ENTER}$. Notice another triangular marker has appeared to mark the right bound. These two markers provide boundaries for the calculator. Finally we are prompted to guess the x-value at which the absolute maximum occurs. Move the cursor near the absolute maximum point or key in an x-value. Press $\boxed{ENTER}$ a third time. We see that the maximum area is 100 square feet when $x = 10$. The rectangle is 10 feet by 10 feet. That makes it a special rectangle. It is actually a square.

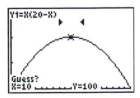   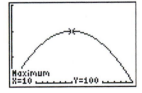

OPTIMIZATION APPLICATION

Section 2.5, Example 2, page 261 *Maximizing Volume* From a thin piece of cardboard 8 inches by 8 inches, square corners are cut out so that the sides can be folded up to make a box. What dimensions will yield a box of maximum volume? What is the maximum volume? (See diagrams in your text.)

The volume of a box = length • width • height. Here we have $V = (8-2x)(8-2x)x$. We will graph this function and find the maximum point on a chosen interval. On the equation-editor screen, clear all functions and turn off the plots. Enter the volume as $Y_1 = (8-2x)(8-2x)x$. Since the original material was 8" by 8" and equal squares are being cut out at the corners and the sides folded up, the domain must be in the closed interval between 0 and 4. Use the window below and then graph the function.

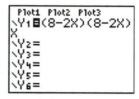 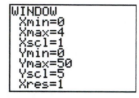 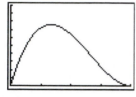

It is worth noting that since this is a cubic function it actually has no absolute maximum, however, we are using it to model a real world problem with constraints and therefore are working on a closed interval. There is an absolute maximum over every closed interval. Here we use "4:maximum" from the CALCULATE menu to find the maximum volume. The following screens are provided to guide you through the process.

 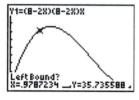 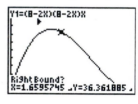 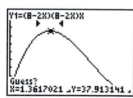

Press ENTER and the maximum point is revealed.

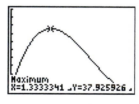

Sometimes the calculator looses some accuracy and we must realize that $x = 1.3333341$ actually means that $x = 1.333333...$ or $x = \dfrac{4}{3}$. To find the other dimensions, go to the home screen by pressing [2nd] [QUIT]. Press [8] [−] [2] [(] [4] [÷] [3] [)] and then press ENTER to see the other dimensions of the box.

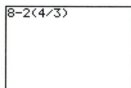 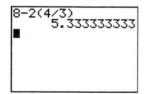

The length and width of the box are equal. They are both $5\frac{1}{3}$ inches. The height is $1\frac{1}{3}$ inches or $\frac{4}{3}$ inches.

To find the volume, press [2nd] [ANS] [×] [2nd] [ANS] [×] [(] [4] [÷] [3] [)]. This takes the previous answer on the home

screen, $5\frac{1}{3}$, and multiplies it by itself and then by the height, $\frac{4}{3}$. The decimal approximation of the volume is given.

   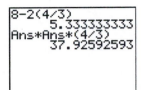

To write this volume as a mixed number, subtract the whole number, 37, and then press [MATH] [ENTER]

[ENTER] to turn the decimal part into a fraction. The volume is $37\frac{25}{27}$ cubic inches.

```
            5.333333333
Ans*Ans*(4/3)
           37.92592593
Ans-37
           .9259259259
Ans▶Frac
                  25/27
■
```

THE LISTS (as a spread sheet)

The lists can be used like a spread sheet to solve problems that involve mathematical formulas.

Section 2.5, Technology Connection, page 268   Use the calculator to fill in the table:

Lot size, x	Number of Recorders, $\dfrac{2500}{x}$	Average Inventory, $\dfrac{x}{2}$	Carrying costs, $10 \cdot \dfrac{x}{2}$	Cost of each order, $20+9x$	Reorder Costs, $(20+9x)\dfrac{2500}{x}$	Total Inventory Costs, $C(x)=10 \cdot \dfrac{x}{2}+(20+9x)\dfrac{2500}{x}$
2500	1	1250	$12,500	$22,200	$22,520	$35,020
1250	2	625	$6,250	$11,270	$22,540	
500	5	250	$2,500	$4,520		
250	10	125				
167	15	84				
125	20					
100	25					
90	28					
50	50					

To fill in the table, press [STAT] [1] or [STAT] [ENTER] to access the statistical list-editor screen. Clear existing entries in the lists as indicated on page 43-44 of this manual. The entries in the first column of the table go in L_1.

```
L1      L2      L3      1          L1      L2      L3      1
2500    ------  ------             250
1250                              167
500                              125
250                              100
167                               90
125                               50
100                               
L1(1)=2500                       L1(10) =
```

Once this is done the spread sheet ability of the calculator can be used to fill in the rest of the table. Now we move the cursor to highlight "L_2" and press ⃞2 ⃞5 ⃞0 ⃞0 ⃞÷ 2nd [L1]. ([L1] is the second function associated with the ⃞1 key.) Then press ENTER and watch L_2 fill in. Notice that two of the entries have decimals. These numbers represent the number of recorders and a decimal makes no sense, therefore we round those entries to the next whole number as in the screen below on the right. Position the cursor over 14.97 and press ⃞1 ⃞5. Position the cursor over 27.778 and press ⃞2 ⃞8.

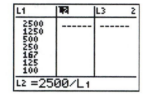

```
L1      L2      L3      2     L1      L2      L3      2     L1      L2      L3      2     L1      L2      L3      2
2500    ------  ------        2500    1       ------       500     5                   500     5
1250                         1250    2                   250     10                  250     10
500                          500     5                   167     14.97               167     15
250                          250     10                  125     20                  125     20
167                          167     14.97               100     25                  100     25
125                          125     20                   90     27.778               90     28
100                          100     25                   50     50                   50     50
L2 =2500/L1                   L2(1)=1                      L2(9) =50                    L2(6) =20
```

Now we move the cursor to highlight "L_3" and press 2nd [L1] ⃞÷ ⃞2. ([L1] is the second function associated with the ⃞1 key.) Then press ENTER and watch L_3 fill in. Again we round up the entries containing decimals.

```
L1      L2      L3      3     L1      L2      L3      3     L1      L2      L3      3     L1      L2      L3      1
2500    1       ------        2500    1       1250        2500    1       1250        250     10      125
1250    2                    1250    2       625         1250    2       625         167     15      84
500     5                    500     5       250         500     5       250         125     20      63
250     10                   250     10      125         250     10      125         100     25      50
167     15                   167     15      83.5        167     15      84          90      28      45
125     20                   125     20      62.5        125     20      63          50      50      25
100     25                   100     25      50          100     25      50          
L3 =L1/2                      L3(1)=1250                   L3(1)=1250                   L1(10) =
```

Now we move the cursor to highlight "L_4" and press ⃞1 ⃞0 ⃞× 2nd [L3]. ([L3] is the second function associated with the ⃞3 key.) Then press ENTER and watch L_4 fill in.

```
L2      L3      L4      4     L2      L3      L4      4     L2      L3      L4      4
1       1250    ------        1       1250    12500       10      125     1250
2       625                  2       625     6250        15      84      840
5       250                  5       250     2500        20      63      630
10      125                  10      125     1250        25      50      500
15      84                   15      84      840         28      45      450
20      63                   20      63      630         50      25      250
25      50                   25      50      500         
L4 =10*L3                     L4(1)=12500                  L4(10) =
```

Now we move the cursor to highlight "L_5" and press ⃞2 ⃞0 ⃞+ ⃞9 ⃞× 2nd [L1]. ([L1] is the second function associated with the ⃞1 key.) Then press ENTER and watch L_5 fill in.

```
L3      L4      L5      5     L3      L4      L5      5     L3      L4      L5      5
1250    12500   ------        1250    12500   22520       125     1250    2270
625     6250                 625     6250    11270       84      840     1523
250     2500                 250     2500    4520        63      630     1145
125     1250                 125     1250    2270        50      500     920
84      840                  84      840     1523        45      450     830
63      630                  63      630     1145        25      250     470
50      500                  50      500     920         
L5 =20+9*L1                   L5(1)=22520                  L5(10) =
```

Now we move the cursor to highlight "L_6" and press 2nd [L5] ⃞× 2nd [L2]. ([L5] and [L2] are the second functions associated with the ⃞5 and ⃞2 keys, respectively.) Then press ENTER and watch L_6 fill in.

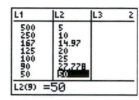
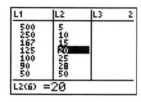

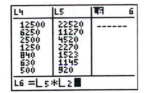

Now we move the cursor to highlight "L_6" and press ▶. This reveals a list with no name. If there is a named list, press DEL to remove it. The calculator is in ALPHABETIC mode as indicated by the ⓐ in the upper right corner of the screen. We name it "COST" by pressing C O S T (C, O, S, and T are the alphabetic characters associated with the PRGM, 7, LN, and 4 keys, respectively.) Since the calculator was in ALPHABETIC mode, we did not have to press the ALPHA before each letter. Press ENTER and the list is named.

With "COST" highlighted, press 2nd [L4] + 2nd [L6]. ([L4] and [L6] are the second functions associated with the 4 and 6 keys, respectively.) Then press ENTER and watch COST fill in.

We search the "COST" list for the smallest number and find that $23,500 is the lowest total inventory cost.

We highlight that element and use the ◀ to return to L_1. Here we find that a lot size of 100 recorders results in the lowest total inventory cost.

Use the values to fill in the table.

The following is an explanation of how to solve the problem graphically.

1.   Graph $C(x) = 10 \cdot \dfrac{x}{2} + (20 + 9x)\dfrac{2500}{x}$ over the interval $[0, 2500]$.

On the equation-editor screen, delete all previous entries and turn off the plots. Then enter the cost equation as $Y_1 = 10 \cdot \dfrac{x}{2} + (20 + 9x)\dfrac{2500}{x}$. An appropriate window is shown on the next page. Press GRAPH after setting the window.

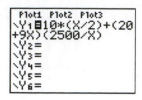 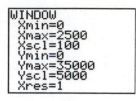 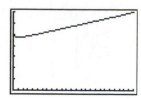

2.   Graphically estimate the minimum value and where it occurs.

Press [2nd] [CALC] [3] to select "3:minimum" from the CALCULATE menu. This feature has been used several times before in this manual. Select a left bound. Select a right bound. Make a guess.

 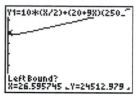 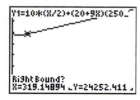 

Once the guess has been made, the minimum appears on the screen.

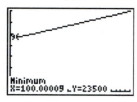

As we found in the table, the minimum total inventory cost is $23,500 and that occurs when the lot size is 100 television sets.

MARGINAL COST, REVENUE, AND PROFIT

Section 2.6, Example 1, page 276 *Business: Marginal Cost, Revenue, and Profit*   Given $C(x) = 62x^2 + 27,500$ and $R(x) = x^3 - 12x^2 + 40x + 10$ find each of the following:

(a)   total profit, $P(x)$.

Profit = Revenue − Cost

$$P(x) = x^3 - 12x^2 + 40x + 10 - (62x^2 + 27500)$$
$$P(x) = x^3 - 12x^2 + 40x + 10 - 62x^2 - 27500$$
$$P(x) = x^3 - 74x^2 + 40x - 27490$$

(b)   total cost, revenue, and profit from the production and sale of 50 units of the product.

On the equation-editor screen, the cost function becomes $Y_1 = 62x^2 + 27,500$ and the revenue function becomes $Y_2 = x^3 - 12x^2 + 40x + 10$. Move the cursor immediately to the right of "Y₃=" and press [VARS] [▶] [1] [2] [−] [VARS] [▶] [1] [1]. Now we have the cost function in the calculator as Y_1, the revenue function as Y_2, and the profit function as Y_3.

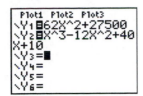 

To evaluate each of the three functions when 50 units are produced and sold, go to the home screen by pressing [2nd] [QUIT]. To find $C(50)$ press [VARS] [▶] [1] [1] [(] [5] [0] [)] [ENTER] and we see $C(50) = \$182,500$. To find $R(50)$ press [VARS] [▶] [1] [2] [(] [5] [0] [)] [ENTER] and we see $R(50) = \$97,010$. To find $P(50)$ press [VARS] [▶] [1] [3] [(] [5] [0] [)] [ENTER] and we see $P(50) = -\$85,490$. Negative profit indicates a loss.

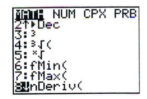 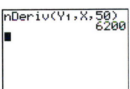 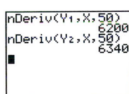

(c)   The marginal cost, revenue, and profit when 50 units are produced and sold. On the home screen, press [MATH] [8] to select "8:nDeriv(" from the MATH submenu in the MATH menu. To find $C'(50)$ press [VARS] [▶] [1] [1] [,] [X,T,Θ,n] [,] [5] [0] [)] [ENTER] and we see that $C'(50) = \$6200$. To find $R'(50)$ press [MATH] [8] [VARS] [▶] [1] [2] [,] [X,T,Θ,n] [,] [5] [0] [)] [ENTER] and we see that $R'(50) = \$6340$. To find $P'(50)$ press [MATH] [8] [VARS] [▶] [1] [3] [,] [X,T,Θ,n] [,] [5] [0] [)] [ENTER] and we see that $P'(50) = \$140$.

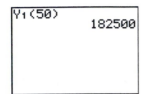 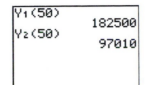 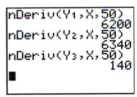

Chapter 3
Exponential and Logarithmic Functions

GRAPHING LOGARITHMIC FUNCTIONS

Section 3.2, Technology Connection, page 320 Here we show two ways to graph $y = \log_2 x$.

First, we use the DRAW INVERSE feature in the DRAW submenu in the DRAW menu. The equation, $y = \log_2 x$, must first be rewritten as the exponential equation $2^y = x$. These equations are different representations of the same equation. On the graphing calculator, we must have an equation solved for y in order to graph it. Therefore, we interchange the x and y in $2^y = x$ and graph $Y_1 = 2^x$. This equation is the inverse of $y = \log_2 x$. With $Y_1 = 2^x$ on the equation-editor screen, press $\boxed{\text{ZOOM}}$ $\boxed{6}$ to graph the function on a standard viewing window. Now press $\boxed{\text{2nd}}$ $[\text{QUIT}]$ to go to the home screen. Press $\boxed{\text{2nd}}$ $[\text{DRAW}]$. ($[\text{DRAW}]$ is the second function associated with the $\boxed{\text{PRGM}}$ key.) Choose "8:DrawInv" by pressing $\boxed{8}$ or by highlighting the 8 and pressing $\boxed{\text{ENTER}}$. Press $\boxed{\text{VARS}}$ $\boxed{\blacktriangleright}$ $\boxed{1}$ $\boxed{1}$ or $\boxed{\text{VARS}}$ $\boxed{\blacktriangleright}$ $\boxed{\text{ENTER}}$ $\boxed{\text{ENTER}}$ to choose Y_1. Press $\boxed{\text{ENTER}}$ again and the graph and its inverse displayed on the standard viewing window. The graph of the inverse of Y_1 is the graph of $y = \log_2 x$.

  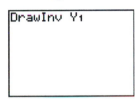 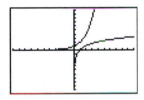

The second way to graph $y = \log_2 x$ is to use the Change-of-Base Formula. See Theorem III on page 321 in your textbook. The Change-of-Base Formula is written here for clarity: $\log_b M = \dfrac{\log_a M}{\log_a b}$. There are two logarithm buttons on the calculator. One is the $\boxed{\text{LOG}}$ key. It represents $\log_{10}$ and is referred to as a common logarithm. The other is the $\boxed{\text{LN}}$ key. It represents $\log_e$ and is referred to as a natural logarithm. Either can be used in the Change-of-Base Formula. The equation $y = \log_2 x$ can be rewritten as $y = \dfrac{\log (x)}{\log (2)}$ or $y = \dfrac{\ln (x)}{\ln (2)}$ using the Change-of-Base Formula.

On the equation-editor screen clear all entries and turn off all plots. Enter $Y_1 = \dfrac{\log (x)}{\log (2)}$ and press $\boxed{\text{ZOOM}}$ $\boxed{6}$ to graph the function on the standard viewing window.

 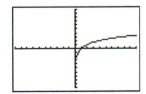

With this graph the TABLE and TRACE features are available. Using the DRAW INVERSE feature, none of the regular features associated with a graph are accessible.

SOLVING EXPONENTIAL EQUATIONS

Section 3.2, Technology Connection, page 325   Solve $e^t = 40$ using a graphing calculator.

Either the INTERSECT feature or the ZERO feature could be used to solve this equation. Here we use the INTERSECT feature. The ZERO feature is discussed on page 20 in this manual. Put the left side of the equation in Y_1 and the right side in Y_2 on the equation-editor screen. With the cursor immediately to the right of "Y_1=" press [2nd] [e^x] [X,T,Θ,n] [ )]. ([e^x] is the second function associated with the [LN] key.) With the cursor immediately to the right of "Y_2=" press [4] [0]. There are many choices for a good viewing window, but the maximum y-value must be at least 40 since we will have a horizontal line at $y = 40$. Once the window is set, press [GRAPH].

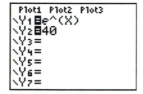   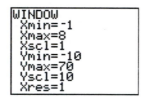   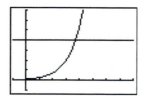

With the graph displayed, press [2nd] [CALC] [5] to choose "5:intersect." The calculator will pose three questions.

The query "First Curve?" appears at the bottom of the screen. The blinking cursor is positioned on the graph of Y_1 as in the screen below on the left. Notice the upper left corner of this screen displays the equation of the curve the cursor is indicating. Press [ENTER] to indicate that this is the first curve involved in the intersection.

Next the query "Second Curve?" appears at the bottom of the screen. The blinking cursor is now positioned on the graph of Y_2 as in the screen below on the right. Notice the upper left corner of this screen displays the equation of the curve the cursor is indicating. Press [ENTER] to indicate that this is the second curve. We identify the curves for the calculator since we could have as many as ten graphs on the screen at once.

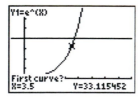   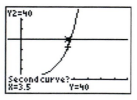

After we identify the second curve, the query "Guess?" appears at the bottom of the screen. Since there is only one point of intersection, press [ENTER] and the point of intersection is displayed.

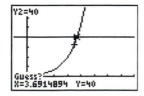   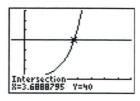

The x-coordinate is the solution to the equation. The solution is $t \approx 3.7$.

EXPONENTIAL REGRESSION

Section 3.3, Technology Connection, page 341   The table below shows data regarding world population growth.

Year	World Population (in billions)
1927	2
1960	3
1974	4
1987	5
1998	6

To find an exponential equation that models the data, enter the data as ordered pairs on the statistical list-editor screen. To clear existing lists press [STAT] [4] [2nd] [L1] [,] [2nd] [L2] [,] [2nd] [L3] [,] [2nd] [L4] [,] [2nd] [L5] [,] [2nd] [L6] [ENTER]. (L_1 through L_6 are the second functions associated with the numeric keys [1] through [6].) The lists can also be cleared by first accessing the statistical list-editor screen by pressing [STAT] [ENTER] or [STAT] [1]. For each list that contains entries, use the arrow keys to move the cursor to highlight the name of the list at the top of the column and press [CLEAR] [▼] or [CLEAR] [ENTER].

Once the lists are cleared, enter the data points. We will enter the number of years in L_1 and the population in L_2. Position the cursor just below "L_1." Press [1] [9] [2] [7] [ENTER]. Continue typing the x-values, 1960, 1974, 1987, and 1998, each followed by [ENTER]. The entries can be followed by [▼] rather than [ENTER] if desired. Press [▶] to move the cursor just below "L_2." Type the population in billions 2, 3, 4, 5, and 6, each followed by [ENTER] or [▼]. Note that the coordinates of each point must be in the same position in both lists.

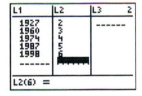

The calculator's EXPONENTIAL REGRESSION feature can be used to fit an exponential equation to the data. With the data points in the lists, press [STAT] [▶] [0] to select "0:ExpReg" from the STATISTICAL CALCULATIONS menu. Now press [2nd] [L1] [,] [2nd] [L2] to let the calculator know which lists contain the data. (Even though the calculator was programmed to assume the x- and y-coordinates of data points are in L_1 and L_2, respectively, it is good practice to specify the lists being used.) Before the regression equation is found, it is possible to select a y-variable to which the equation will be stored on the equation-editor screen. To do this, press [,] [VARS] [▶] [ENTER] [ENTER]. Finally to display the coefficients of the regression equation press [ENTER] again.

 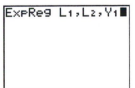 

With the regression equation in Y_1, we evaluate $Y_1(2008)$. This predicts the world's population in 2008.

On the home screen, press VARS ▶ ENTER ENTER (2 0 0 8) ENTER.

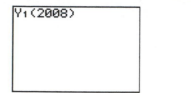

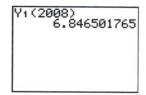

In 2008, the world's population will be approximately 6.8465 billion.

EXPONENTIAL DECAY

Section 3.4, Example 1, page 353 *Life Science: Decay* Strontium-90 has a decay rate of 2.8% per year. The rate of change of an amount N is given by $\dfrac{dN}{dt} = -0.028N$.

(a)   Find the function that satisfies the equation. Let N_0 represent the amount present at $t = 0$.

Since $\dfrac{dP}{dt} = -kP$ and we have $\dfrac{dN}{dt} = -0.028N$, we know $k = -0.028$. Using the function $P(t) = P_0 e^{-kt}$ and substituting, we have $N(t) = N_0 e^{-0.028t}$.

(b)   Suppose that 1000 grams of strontium-90 is present at $t = 0$. How much will remain after 70 years?

The function in part (a) is now $N(t) = 1000e^{-0.028t}$. On the equation-editor screen enter $Y_1 = 1000e^{-0.028x}$.

The t in the function is x in the calculator. Remember $[e^x]$ is the second function associated with the LN key. Press 2nd [QUIT] to go to the home screen. To evaluate the function when $t = 70$ press VARS ▶ ENTER ENTER (7 0) ENTER.

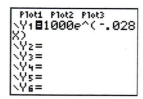

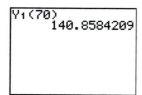

After 70 years, about 140.9 grams of strontium-90 remains.

(c)   After how long will half of the 1000 grams remain? (Here we are finding the half-life of the substance.) With $Y_1 = 1000e^{-0.028x}$ already on the equation-editor screen, use $Y_2 = 500$ as the second equation. The x-coordinate of the intersection point will be the answer to the question.

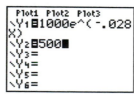

The *x*-axis is the time axis and therefore should not have any negative values. The *y*-axis is the axis which gives the quantity of the substance, so it should not be negative either. The answer is in first quadrant. Trial and error will be necessary to find a good window. The one below works well. Once we have the desired window, press GRAPH to see the function and the horizontal line displayed.

 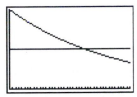

With the graph displayed, press 2nd [CALC] 5 to choose "5:intersect." The calculator will pose three questions.

The query "First Curve?" appears at the bottom of the screen. The blinking cursor is positioned on the graph of Y_1 as in the screen below on the left. Notice the upper left corner of this screen displays the equation of the curve the cursor is indicating. Press ENTER to indicate that this is the first curve involved in the intersection.

Next the query "Second Curve?" appears at the bottom of the screen. The blinking cursor is now positioned on the graph of Y_2 as in the screen below on the right. Notice the upper left corner of this screen displays the equation of the curve the cursor is indicating. Press ENTER to indicate that this is the second curve.

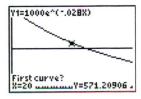 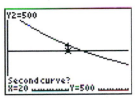

After we identify the second curve, the query "Guess?" appears at the bottom of the screen. Since there is only one point of intersection, press ENTER and the point of intersection is displayed.

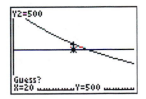 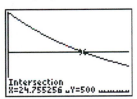

The *x*-coordinate is the solution to the equation, $t \approx 24.755$ years. We can write this answer in years, months, and days by immediately going to the home screen. Press 2nd [QUIT] to do this. Once on the home screen press X,T,Θ,n ENTER to see the answer revealed. We subtract the years away by pressing − 2 4 ENTER. Notice that "Ans" is displayed when − is pressed.

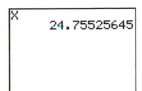 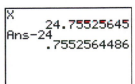

The number that remains is the decimal part of a year. To convert it to months, multiply by 12 by pressing ⊠ ① ② ⟦ENTER⟧. "Ans" is displayed again. We see that the decimal converted to 9 months.

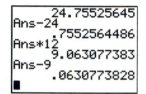

Subtract away the months by pressing ⊟ ⑨ ⟦ENTER⟧. The decimal that remains is the decimal part of a month. To convert this to days, we need to multiply it by the number of days in a month. Since the number of days in a month varies, use 30 as a close approximation. Multiply by 30 by pressing ⊠ ③ ⓪ ⟦ENTER⟧. The decimal converts to approximately 2 days.

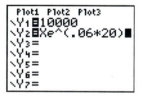 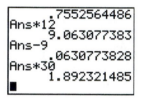

The half-life of the substance is 24 years, 9 months, and 2 days.

PRESENT VALUE

Section 3.4, Example 4, page 356 *Business: Present Value* Following the birth of their grand daughter, two grandparents want to make an initial investment of P_0 that will grow to $10,000 by the child's 20th birthday. Interest is compounded continuously at 6%. What should the initial investment be?

Using the formula for continuous compounding, $P = P_0 e^{kt}$, we have $10,000 = P_0 e^{0.06 \cdot 20}$. This can be solved using the INTERSECT feature of the CALCULATE menu by putting the left side of the equation in Y_1 and the right side in Y_2. P_0, the present value, becomes x in the calculator. Remember $[e^x]$ is the second function associated with the ⟦LN⟧ key. The 6% interest goes into the formula where the "k" was. Remember $6\% = .06$. The "t" in the formula is time and it is 20 years.

```
Plot1  Plot2  Plot3
\Y1■10000
\Y2■Xe^(.06*20)■
\Y3=
\Y4=
\Y5=
\Y6=
\Y7=
```

The *x*-axis represents the present value and the *y*-axis is the future value on the calculator. Negative values would make no sense for either. This means we need to only look in first quadrant. Trial and error will be necessary to find a good window. One good window choice is shown below. Once we have the desired window, press ⟦GRAPH⟧ to see the function and the horizontal line displayed.

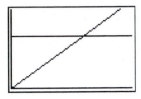

With the graph displayed, press [2nd] [CALC] [5] to choose "5:intersect." The calculator will pose three questions.

The query "First Curve?" appears at the bottom of the screen. The blinking cursor is positioned on the graph of Y_1 as in the screen below on the left. Notice the upper left corner of this screen displays the equation of the curve the cursor is indicating. Press [ENTER] to indicate that this is the first curve involved in the intersection.

Next the query "Second Curve?" appears at the bottom of the screen. The blinking cursor is now positioned on the graph of Y_2 as in the screen below on the right. Notice the upper left corner of this screen displays the equation of the curve the cursor is indicating. Press [ENTER] to indicate that this is the second curve.

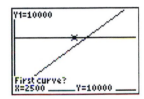
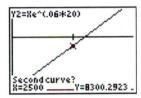

After we identify the second curve, the query "Guess?" appears at the bottom of the screen. Since there is only one point of intersection, press [ENTER] and the point of intersection is displayed.

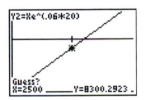
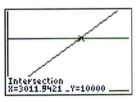

The x-coordinate is the solution to the equation. The grandparents need to deposit $3011.94 at the child's birth for it to grow to $10,000 by the child's 20th birthday.

Chapter 4
Integration

SUMMATION NOTATION, $\sum$

Summations can be evaluated using the SUMMATION feature from the MATH submenu of the LIST menu and the SEQUENCE feature from the OPERATIONS submenu of the LIST menu.

Section 4.1, Example 5, page 395 Express $\sum_{i=1}^{4} 3^i$ without using summation notation. Here we evaluate $\sum_{i=1}^{4} 3^i$ using the graphing calculator. The commands are "sum(seq(" followed by the sequence, the variable name, the beginning number, the ending number, and how to increment. Then the two pairs of parentheses are closed.

Press 2nd [LIST]. ([LIST] is the second function associated with the STAT key.) Press the ▶ key twice to access the MATH submenu of the LIST menu. Select "5:sum(" by pressing 5.

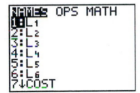  

Press 2nd [LIST] ▶ to access the OPERATIONS submenu of the LIST menu. Here select "5:seq(" by pressing 5. Press 3 ^ ALPHA I. (I is the alphabetic character associated with the x^2 key.) With the sequence defined, press , ALPHA I to name the variable.

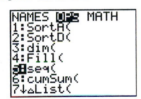  

Press , 1 , 4. This tells the calculator where to start and where to stop. Press , 1)) to tell the calculator how to count and to close both sets of parenthesis. Press ENTER and the answer is revealed. $\sum_{i=1}^{4} 3^i = 120$.

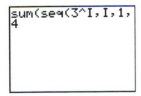 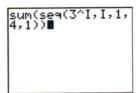 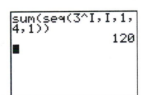

THE fnInt FEATURE

Definite integrals can be evaluated on the home screen using the FUNCTION INTEGRAL feature, "fnInt(," from the MATH submenu of the MATH menu.

Section 4.3, Technology Connection, page 418 Evaluate $\int_{-1}^{2}(-x^3+3x-1)\,dx$ using the FUNCTION INTEGRAL

feature on the home screen.

First select "9:fnInt(" from the MATH submenu of the MATH menu. Then enter the function, the variable, and the lower and upper limits of integration. Press [MATH] [9] [(-)] [X,T,Θ,n] [^] [3] [+] [3] [X,T,Θ,n] [-] [1] [,] [X,T,Θ,n] [,] [(-)]

[1] [,] [2] [)] [ENTER]. We find that $\int_{-1}^{2}(-x^3+3x-1)\,dx = -2.25$.

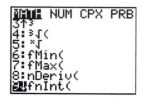

If the function has been entered on the equation-editor screen, as $Y_1 = -x^3 + 3x - 1$ we can evaluate it by

entering fnInt (Y_1, X, -1, 2) on the home screen. Press [MATH] [9] [VARS] [▶] [ENTER] [ENTER] [,] [X,T,Θ,n] [,] [(-)] [1] [,] [2] [)]

[ENTER].

THE $\int f(x)\,dx$ FEATURE

Definite integrals can also be evaluated on the graph screen using the NUMERICAL INTEGRAL feature,

" $\int f(x)\,dx$," from the CALCULATE menu.

Section 4.3, Technology Connection, page 418 Evaluate $\int_{-1}^{2}(-x^3+3x-1)\,dx$ on the graph screen using the

NUMERICAL INTEGRAL feature.

Graph $Y_1 = -x^3 + 3x - 1$ in a window that contains the interval [-1, 2]. We will use [-3, 3, -6, 6]. Then

select the "7: $\int f(x)\,dx$" from the CALCULATE menu by pressing [2nd] [CALC] [7].

 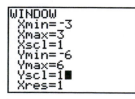 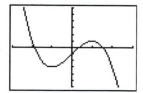 

We are prompted to enter the lower limit of integration. Press [(-)] [1]. Then press [ENTER]. Now enter the

upper limit of integration by pressing [2]. Press [ENTER] again. The calculator shades the area above and below the

curve on [-1, 2] and returns the value of the definite integral on this interval.

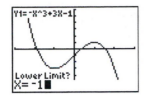 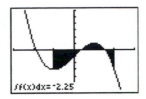 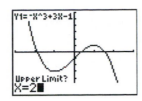

THE fnInt FEATURE; AREA BETWEEN TWO CURVES

The area between two curves can be found using the INTERSECT feature from the CALCULATE menu and the FUNCTION INTEGRAL feature from the MATH submenu of the MATH menu.

Section 4.4, Technology Connection, page 429 Find the area bounded by the two graphs $Y_1 = -2x - 7$ and $Y_2 = -x^2 - 4$. First graph both functions and determine which function is the upper and which is the lower graph. The equation-editor screen, the window, and the graph are displayed below.

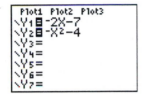 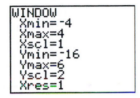 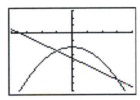

To determine which curve is the upper one, press [TRACE]. A flashing cursor will appear on Y_1 and its equation will be displayed in the upper left corner of the screen. Press either the up- or down-arrow key to move the cursor to Y_2 and its equation will be displayed. From examining these screens we see that Y_2 is the upper curve for the region between the two intersection points.

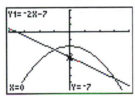 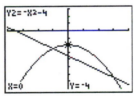

Go to the home screen by pressing [2nd] [QUIT]. Select the "9:fnInt(" by pressing [MATH] [9]. To enter the function, enter the upper curve, Y_2, minus the lower curve, Y_1. Then enter the variable. Press [VARS] [▶] [1] [2] [−] [VARS] [▶] [1] [1] [,] [X,T,Θ,n] [,].

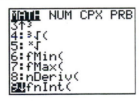 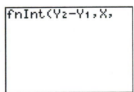

To find the lower and upper limits of integration, return to the graph screen by pressing [GRAPH]. Then use the INTERSECT feature from the CALCULATE menu to find the points of intersection. The x-coordinates of the points are the lower and upper limits of integration. With the graph displayed, press [2nd] [CALC] [5] to choose "5:intersect." The calculator will pose three questions.

The query "First Curve?" appears at the bottom of the screen. The blinking cursor is positioned on the graph of Y_1 as in the screen below on the left. Notice the upper left corner of this screen displays the equation of the curve the cursor is indicating. Press [ENTER] to indicate that this is the first curve involved in the intersection.

Next the query "Second Curve?" appears at the bottom of the screen. The blinking cursor is now positioned on the graph of Y_2 as in the screen below on the right. Notice the upper left corner of this screen displays the equation of the curve the cursor is indicating. Press [ENTER] to indicate that this is the second curve.

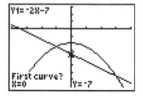   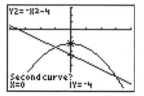

After we identify the second curve, the query "Guess?" appears at the bottom of the screen. Press [ENTER] and the point of intersection nearest the cursor is displayed.

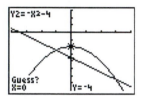   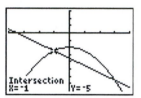

This x-coordinate is the lower limit of integration. So the lower limit of integration is -1.

We repeat the process to find the upper limit of integration, which is the x-coordinate of the other point of intersection. Press [2nd] [CALC] [5] to choose "5:intersect." The calculator will pose three questions.

The query "First Curve?" appears at the bottom of the screen. The blinking cursor is positioned on the graph of Y_1 as in the screen below on the left. Notice the upper left corner of this screen displays the equation of the curve the cursor is indicating. Press [ENTER] to indicate that this is the first curve involved in the intersection.

Next the query "Second Curve?" appears at the bottom of the screen. The blinking cursor is now positioned on the graph of Y_2 as in the screen below on the right. Notice the upper left corner of this screen displays the equation of the curve the cursor is indicating. Press [ENTER] to indicate that this is the second curve.

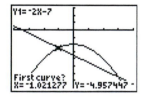   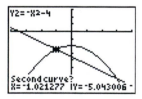

After we identify the second curve, the query "Guess?" appears at the bottom of the screen. This time move the cursor near the point of intersection in the fourth quadrant, since that is the one we have not found. When the cursor is near it, press [ENTER].

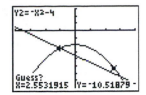   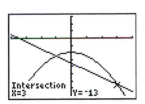

This *x*-coordinate is the upper limit of integration. So the upper limit of integration is 3. Press [2nd] [QUIT] to

return to the home screen. The calculator should be waiting for the lower and upper limits of integration. Press [(-)]

[1] [,] [3] [)]. Then press [ENTER]. To express the answer as a fraction, press [MATH] [ENTER] [ENTER].

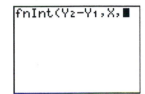

The area between the two curves is $10\frac{2}{3}$ square units or $\frac{32}{3}$ square units.

AVERAGE VALUE

Section 4.4, Example 5, page 432   Find the average value of $f(x) = x^2$ over the interval [0, 2].

The formula to find the average value of a function over a closed interval is given by: $y_{av} = \frac{1}{b-a}\int_a^b f(x)$.

For this example $a = 0$ and $b = 2$. Go to the home screen and press [(] [1] [÷] [(] [2] [−] [0] [)] [)] [MATH] [9] [X,T,Θ,n] [x²]

[,] [X,T,Θ,n] [,] [0] [,] [2] [)]. Press [ENTER] to find the average value over the interval. To convert the answer into a

fraction, press [MATH] [ENTER] [ENTER].

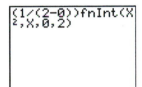   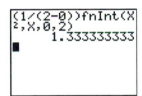   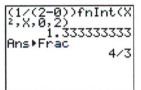

The average value of $f(x) = x^2$ over the interval [0, 2] is $\frac{4}{3}$.

Chapter 5
Applications of Integration

AN ECONOMICS APPLICATION: CONSUMER and PRODUCER SURPLUS

Section 5.1, Example 3, page 474   Given $D(x) = (x-5)^2$ and $S(x) = x^2 + x + 3$,

(a)   find the equilibrium point.

To find the equilibrium point, graph both the supply and demand curves in an appropriate viewing window.

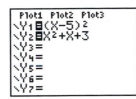 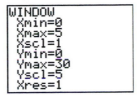 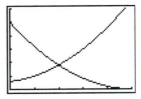

Use the INTERSECT feature from the CALCULATE menu to find the equilibrium point. Press

[2nd] [CALC] [5] to choose "5:intersect" and respond to the three questions "First Curve?," "Second Curve?," and

"Guess?" by pressing [ENTER] three times.

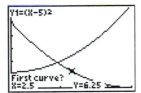 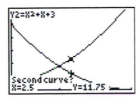 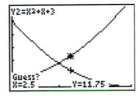 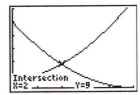

The equilibrium point occurs at (2, $9). The equilibrium quantity is 2 items and the equilibrium price is $9.

(b)   find the consumer surplus at the equilibrium point.

The region between the two curves, the y-axis, and the equilibrium point contains the area that represents both the producer surplus and consumer surplus. First separate the region into two parts by graphing $Y_3 = 9$. This will put a horizontal line through the equilibrium point.

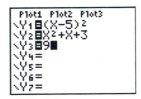   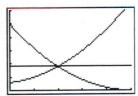

The upper region represents the consumer surplus as is illustrated on the screen below. This screen is provided to clarify which area we are finding. Instructions to shade the region are not given here.

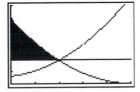

To find the upper and lower curves before integrating, press [TRACE] then use the [◄] key to move the cursor to part of the curve that serves as a boundary of the region in question. We see that Y_1 is the upper boundary. We use the ▲ and ▼ keys to move from curve to curve. Y_3 is the lower boundary of this region. Notice the upper left corner of the screen has the equation of the curve the cursor is on.

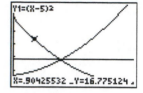 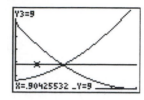 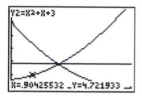

To find the consumer surplus, go to the home screen by pressing [2nd] [QUIT]. Press [MATH] [9] to select "9:fnInt(." Since Y_1 is the upper boundary of the region and Y_3 is the lower boundary, press

[VARS] [▶] [1] [1] [−] [VARS]

[▶] [1] [3] [,] [X,T,Θ,n] [,] to enter the upper boundary minus the lower boundary and the variable name. To integrate from the y-axis to the equilibrium point, press [0] [,] [2] [)]. Then press [ENTER]. The consumer surplus is $14.67.

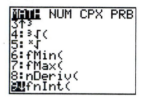 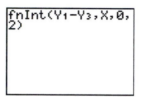 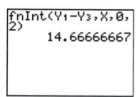

(c)   find the producer surplus at the equilibrium point.

The region between the two curves, the y-axis, and the equilibrium point contains the area that represents both the producer surplus and consumer surplus. We have separated the region into two parts by graphing $Y_3 = 9$.

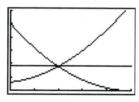

The lower region represents the producer surplus as is illustrated on the screen below. This screen is provided to clarify which area we are finding. Instructions to shade the region are not given here.

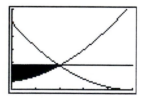

To find the upper and lower curves before integrating, press [TRACE] then use the ◀ key to move the cursor to part of the curve that serves as a boundary of the region in question. We see that Y_3 is the upper boundary. We use the ▲ and ▼ keys to move from curve to curve. Y_2 is the lower boundary of this region. Notice the upper left corner of the screen has the equation of the curve the cursor is on.

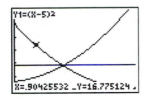
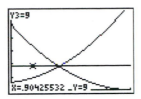
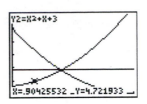

To find the producer surplus, go to the home screen by pressing [2nd] [QUIT]. Press [MATH] [9] to select "9:fnInt(." Since Y_3 is the upper boundary of the region and Y_2 is the lower, press [VARS] [▶] [1] [3] [–] [VARS] [▶] [1] [2] [,] [X,T,Θ,n] [,] to enter the upper boundary minus the lower boundary and the variable name. To integrate from the y-axis to the equilibrium point, press [0] [,] [2] [)]. Then press [ENTER]. The producer surplus is $7.33.

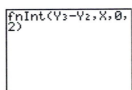
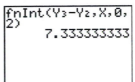

ACCUMULATED PRESENT VALUE

Section 5.2, Example 7, page 482 *Business: Accumulated Present Value* Find the accumulated present value of an investment over a 5-year period if there is a continuous money flow of $2400 per year and the interest rate is 14%, compounded continuously.

The accumulated present value is $\int_0^5 \$2400 e^{-0.14t}\,dt$. This problem can be done on a graphing calculator using the NUMERICAL INTEGRAL feature. On the home screen, press [MATH] [9] [2] [4] [0] [0] [2nd] [e^x] [(-)] [.] [1] [4] [ALPHA] T [)] [,] [ALPHA] T [,] [0] [,] [5] [)]. ([e^x] is the second function associated with the [LN] key and T is the alphabetic character associated with the [4] key.) Press [ENTER] to see that the accumulated present value is $8629.97.

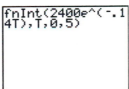
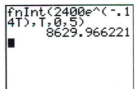

STATISTICS

Section 5.5, Example 4, page 511 The weights w of the students in a calculus class are normally distributed with a mean of 150 pounds and a standard deviation of 25 pounds. Find the probability that a student's weight is from 160 to 180 pounds.

It is not necessary to standardize the weights when using the TI-83 Plus or TI-84 Plus graphing calculator. The calculator will graph the normal density function for the given mean and standard deviation, shade the area between 160 and 180, and express the probability as the area of the shaded region. A fair amount of trial and error might be required to find a suitable window. One good choice is [0, 300, -0.002, 0.02], Xscl = 50, Yscl = 0.01.

Enter these dimensions on the window screen. Clear any entries that are present on the equation-editor screen and be sure the plots are turned off. Then go to the home screen and press [2nd] [DISTR] [▶] [1] or [2nd] [DISTR] [▶] [ENTER] to copy "1:ShadeNorm(" to the home screen. ([DISTR] is the second operation associated with the [VARS] key.) Then enter the left and right endpoints of the interval, the mean, and the standard deviation. To do this, press [1] [6] [0] [,] [1] [8] [0] [,] [1] [5] [0] [,] [2] [5] [)]. Press [ENTER] to see the shaded area.

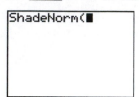  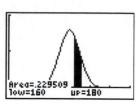

The shaded area is 0.229509, so the probability is approximately 23%.

Chapter 6
Functions of Several Variables

PARTIAL DERIVATIVES

Section 6.2, Technology Connection, page 550 Given the function $f(x, y) = 3x^3 y + 2xy$, use a graphing

calculator that finds derivatives of functions of one variable to find $f_x(-4,1)$ and $f_y(2,6)$.

To find $f_x(-4,1)$, first find $f(x,1)$:

$$f(x, y) = 3x^3 y + 2xy$$
$$f(x,1) = 3x^3(1) + 2x(1)$$
$$= 3x^3 + 2x$$

Now we have a function with one variable, so we use "nDeriv(" from the MATH menu on the home screen

or " dy/dx " from the CALCULATE menu on the graph screen to find the value of the derivative of this function

when $x = -4$.

Enter the function on the equation-editor screen. Press [2nd] [QUIT] to go to the home screen. Press [MATH] [8]

to choose "8:nDeriv(." Then press [VARS] [▶] [1] [1] [,] [X,T,Θ,n] [,] [(-)] [4] [)] [ENTER].

 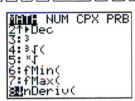 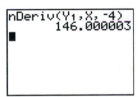

The procedures the calculator uses to calculate the derivative do not always yield an exact answer. Note

that the exact answer is 146, but the calculator produced 146.000003.

With the equation on the equation-editor screen, press [ZOOM] [6] to see the graph displayed in a standard

viewing window. Press [2nd] [CALC] [6] to choose "6: dy/dx " from the CALCULATE menu. Press [(-)] [4] [ENTER].

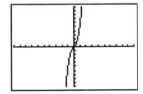  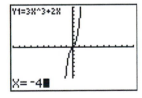 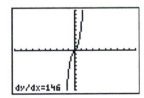

This procedure yields the exact answer.

To find $f_y(2,6)$, first find $f(2, y)$:

$$f(x, y) = 3x^3 y + 2xy$$
$$f(2, y) = 3(2)^3 y + 2(2) y$$
$$= 24y + 4y$$
$$= 28y$$

Now find the derivative of $f(y) = 28y$ when $y = 6$ using "nDeriv(" from the MATH menu. On the home screen, press $\boxed{\text{MATH}}$ $\boxed{8}$ to choose "8:nDeriv(" from the MATH menu. Press $\boxed{2}$ $\boxed{8}$ $\boxed{\text{ALPHA}}$ Y $\boxed{,}$ $\boxed{\text{ALPHA}}$ Y $\boxed{,}$ $\boxed{6}$ $\boxed{)}$ $\boxed{\text{ENTER}}$. (Y is the alphabetic character associated with the $\boxed{1}$ key.)

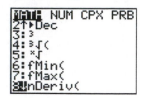   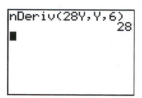

We can also replace y with x and find the derivative of $f(x) = 28x$ when $x = 6$ using "$6: dy/dx$" from the CALCULATE menu. Enter the function on the equation-editor screen. Change your window values to the ones below and press $\boxed{\text{GRAPH}}$. With the graph displayed, press $\boxed{\text{2nd}}$ $[\text{CALC}]$ $\boxed{6}$ $\boxed{6}$ $\boxed{\text{ENTER}}$.

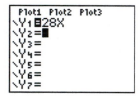   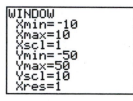   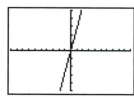   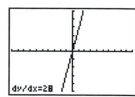

LINEAR REGRESSION

We can use the LINEAR REGRESSION feature in the STATISTICAL CALCULATIONS menu to fit a linear equation to a set of data points.

Section 6.4, page 566 Using the data in the table find the line of best fit. The least-squares technique is shown in the text.

The following table shows the yearly revenue for a rental car company that rents hybrid cars.

Years (in 5 year increments), x	1. 1991	2. 1996	3. 2001	4. 2006	5. 2011
Yearly Revenue (in millions), y	\$5.2	\$8.9	\$11.7	\$16.8	\$

(a)   Make a scatter plot of the data and determine whether the data seem to fit a linear function.

We enter the data as ordered pairs on the statistical list-editor screen. To clear existing lists press $\boxed{\text{STAT}}$ $\boxed{4}$ $\boxed{\text{2nd}}$ $[\text{L1}]$ $\boxed{,}$ $\boxed{\text{2nd}}$ $[\text{L2}]$ $\boxed{,}$ $\boxed{\text{2nd}}$ $[\text{L3}]$ $\boxed{,}$ $\boxed{\text{2nd}}$ $[\text{L4}]$ $\boxed{,}$ $\boxed{\text{2nd}}$ $[\text{L5}]$ $\boxed{,}$ $\boxed{\text{2nd}}$ $[\text{L6}]$ $\boxed{\text{ENTER}}$. (L_1 through L_6 are the second functions associated with the numeric keys $\boxed{1}$ through $\boxed{6}$.) The lists can also be cleared by first accessing the statistical list-editor screen by pressing $\boxed{\text{STAT}}$ $\boxed{\text{ENTER}}$ or $\boxed{\text{STAT}}$ $\boxed{1}$ to select "1:Edit" from the menu. For each list that contains entries, use the arrow keys to move the cursor and highlight the name of the list at the top of the column and press $\boxed{\text{CLEAR}}$ $\boxed{\blacktriangledown}$ or $\boxed{\text{CLEAR}}$ $\boxed{\text{ENTER}}$.

Once the lists are cleared, enter the data points. Enter the number of years in L_1 and the revenue amounts in L_2. Position the cursor at the top of column L_1 just below the L_1 heading. Press $\boxed{1}$ $\boxed{\text{ENTER}}$. Continue typing the x-values 2 through 4, each followed by $\boxed{\text{ENTER}}$. The entries can be followed by $\boxed{\blacktriangledown}$ rather than $\boxed{\text{ENTER}}$ if desired. Press

▶ to move to the top of column L_2. Type the revenue amounts, 5.2, 8.9, and so on in succession, each followed by ENTER or ▼. Note that the coordinates of each point must be in the same position in both lists.

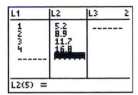

To plot the data points, turn on the STATISTICAL PLOT feature. To access the statistical plot screen, press 2nd [STAT PLOT]. ([STAT PLOT] is the second function associated with the Y= key in the upper left-hand corner of the keypad.)

We will use Plot 1. Since it is already highlighted, we access it by pressing ENTER or 1. The cursor should be positioned over "On" and it should be flashing. Press ENTER to turn on Plot 1. The entries Type, Xlist, and Ylist should be as shown below. The last item, Mark, allows us to choose a box, a cross, or a dot for each point. Here we have selected a box. To select Type and Mark, position the cursor over the appropriate selection and press ENTER. Use the L_1 and L_2 keys (associated with the 1 and 2 numeric keys) to select Xlist and Ylist.

The plot can also be turned on from the equation-editor screen. Press Y= to go to the equation-editor screen. Then assuming that Plot 1 has not yet been turned on and that the desired settings are currently entered for Plot 1 on the statistical plot screen, position the cursor over "Plot 1" and press ENTER. "Plot 1" will be highlighted. Before viewing the plot any existing entries on the equation-editor screen should be cleared. The easiest way to choose a viewing window is to use the ZOOMSTAT feature. Press ZOOM 9 or ZOOM and the down-arrow until the selection cursor is beside "9:ZoomStat" and press ENTER. ZOOMSTAT automatically defines a viewing window that displays all of the data points.

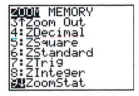
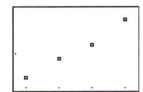

With all the data points in the viewing window it is easy to press WINDOW and modify the window to suit your needs. Below the window has been changed and the new graph is displayed.

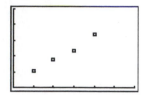

The scatter plot shows that the data is approximately linear.

(b)   Find a linear function that (approximately) fits the data.

The calculator's LINEAR REGRESSION feature can be used to fit a linear equation to the data. With the data points in the lists, press STAT ▶ 4 to select "4:LinReg($ax + b$)" from the STATISTICAL CALCULATIONS menu. Press 2nd [L1] , 2nd [L2] to let the calculator know which lists contain the data. (Even though the calculator was programmed to assume the x- and y-coordinates of data points are in L_1 and L_2, respectively, it is good practice to specify the lists being used.) Before the regression equation is found, it is possible to select a y-variable to which it will be stored on the equation-editor screen. To do this, press , VARS ▶ ENTER ENTER. Finally, to display the coefficients of the regression equation press ENTER again. If the diagnostics have been turned on in your calculator, values of r^2 and r will also be displayed. The correlation coefficient is r. The correlation coefficient is used to describe the strength of the linear relationship between the data points. The closer r is to 1, the better the correlation. The coefficient of determination is r^2.

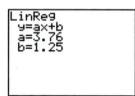
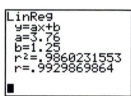

The line of best fit is $y = 3.76x + 1.25$.

If you wish to select DIAGNOSTIC ON mode, press 2nd [CATALOG] and use ▼ to position the triangular selection cursor beside "DiagnosticOn." To alleviate the tedium of scrolling through many items to reach "DiagnosticOn," press D after pressing 2nd [CATALOG] to move quickly to the first catalog item that begins with the letter D. (D is the alphabetic character associated with the x^{-1} key.) Then use ▼ to scroll to "DiagnosticOn." Note that it is not necessary to press ALPHA D. Press ENTER to paste this instruction on the home screen and then press ENTER a second time to set the mode. To select DIAGNOSTIC OFF mode, press 2nd [CATALOG], position the selection cursor beside "DiagnosticOff," press ENTER to paste the instruction on the home screen, and then press ENTER again to set the mode.

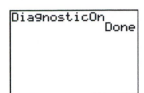

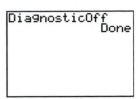

Press GRAPH to see the line displayed with the data points.

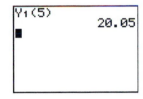

(c)   Use the model to predict the yearly revenue in 2011.

To predict the yearly revenue in 2011, evaluate the regression equation for $x = 5$. (2011 is 20 years after 1991 and in this example we assigned 1 to 1991 and are counting in 5 year increments.) Use any of the methods for evaluating a function presented earlier in this manual. We will use function notation on the home screen. Press

VARS ▶ ENTER ENTER (5) ENTER.

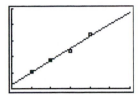

When $x = 5$, $y \approx 20.05$ so we predict the 2011 yearly revenue from renting hybrid cars will be approximately \$20.05 million.

USING 2-VARIABLE STATISTICS

We can use the 2-VARIABLE STATISTICS feature in the STATISTICAL CALCULATIONS menu to fit a linear equation to a set of data points.

Section 6.4, page 569 Using the data in the table find the line of best fit using the 2-VARIABLE STATISTICS feature, and the lists.

The table shows the yearly revenue for a rental car company which rents hybrid cars.

Years (in 5 year increments), x	1. 1991	2. 1996	3. 2001	4. 2006	5. 2011
Yearly Revenue (in millions), y	\$5.2	\$8.9	\$11.7	\$16.8	\$

Press STAT ENTER and enter the data into L_1 and L_2.

We wish to fill in the table and find the regression line.

Variable	c_i	d_i	$c_i - \bar{x}$	$(c_i - \bar{x})^2$	$d_i - \bar{y}$	$(c_i - \bar{x})(d_i - \bar{y})$
Calculator location	L_1	L_2	$L_3 = L_1 - \bar{x}$	$L_4 = L_3{}^2$	$L_5 = L_2 - \bar{y}$	$L_6 = L_3 \cdot L_5$
	1	5.2				
	2	8.9				
	3	11.7				
	4	16.8				
	$\sum_{i=1}^{4} c_i =$	$\sum_{i=1}^{4} d_i =$		$\sum_{i=1}^{4} (c_i - \bar{x})^2 =$		$\sum_{i=1}^{4} (c_i - \bar{x})(d_i - \bar{y}) =$
	$\bar{x} =$	$\bar{y} =$				

Press $\boxed{\text{STAT}}$ $\boxed{\blacktriangleright}$ $\boxed{2}$ to choose "2:2-Var Stats." Now press $\boxed{\text{2nd}}$ $\boxed{\text{L1}}$ $\boxed{,}$ $\boxed{\text{2nd}}$ $\boxed{\text{L2}}$ to tell the calculator where the data is located. ($\boxed{\text{L1}}$ and $\boxed{\text{L2}}$ are the second functions associated with the $\boxed{1}$ and $\boxed{2}$ keys, respectively.) Press $\boxed{\text{ENTER}}$ to reveal the 2-VARIABLE STATISTICS. Use the up- and down-arrow keys to scroll through the screen.

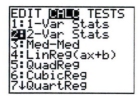 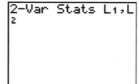 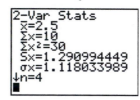 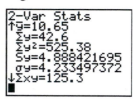

So far we have $\bar{x} = 2.5$, $\bar{y} = 10.65$, $\sum_{i=1}^{4} c_i = 10$, and $\sum_{i=1}^{4} d_i = 42.6$. To find $c_i - \bar{x}$, return to the statistical list-editor screen by pressing $\boxed{\text{STAT}}$ $\boxed{\text{ENTER}}$. Move the cursor to the top of the L_3 list. "L_3" should be highlighted. Press $\boxed{\text{2nd}}$ $\boxed{\text{L1}}$ $\boxed{-}$ $\boxed{\text{VARS}}$ $\boxed{5}$ $\boxed{2}$ $\boxed{\text{ENTER}}$. The keystrokes took us to the VARIABLES menu where we selected "5:Statistics" and then "2: $\bar{x}$."

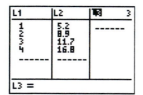 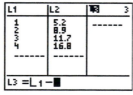  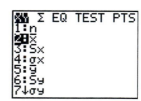

Now we have $c_i - \bar{x}$ in L_3. To find $(c_i - \bar{x})^2$, press $\boxed{\text{STAT}}$ $\boxed{\text{ENTER}}$ and move the cursor to the top of the L_4 list. "L_4" should be highlighted. Press $\boxed{\text{2nd}}$ $\boxed{\text{L3}}$ $\boxed{x^2}$ $\boxed{\text{ENTER}}$.

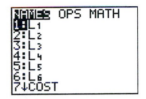  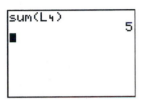

Now we have $(c_i - \overline{x})^2$ in L_4. To find $\sum_{i=1}^{4}(c_i - \overline{x})^2 =$, go to the home screen by pressing [2nd] [QUIT]. Press

[2nd] [LIST]. ([LIST] is the second function associated with the [STAT] key.) Press [▶] [▶] to reach the MATH submenu of

the LIST menu. Press [5] to choose "5:sum(." Now press [2nd] [L4] [)] [ENTER]. ([L4] is the second function associated

with the [4] key.) Now we know $\sum_{i=1}^{4}(c_i - \overline{x})^2 = 5$.

To find $d_i - \overline{y}$, return to the statistical list-editor screen by pressing [STAT] [ENTER]. Move the cursor to the

top of the L_5 list. "L_5" should be highlighted. Press [2nd] [L2] [−] [VARS] [5] [5] [ENTER]. The keystrokes took us to the

VARIABLES menu where we selected "5:Statistics" and then "5: $\overline{y}$."

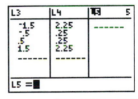 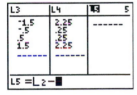  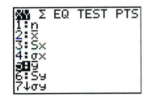

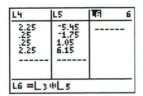 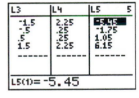

Now we have $d_i - \overline{y}$ in L_5. To find $(c_i - \overline{x})(d_i - \overline{y})$, move the cursor to the top of the L_6 list. "L_6" should

be highlighted. Press [2nd] [L3] [×] [2nd] [L5] [ENTER].

Now we have $(c_i - \overline{x})(d_i - \overline{y})$ in L_6. To find $\sum_{i=1}^{4}(c_i - \overline{x})(d_i - \overline{y}) =$ go to the home screen by pressing [2nd]

[QUIT]. Press [2nd] [LIST]. ([LIST] is the second function associated with the [STAT] key.) Press [▶] [▶] to reach the MATH submenu of the LIST menu. Press [5] to choose "5:sum." Now press [2nd] [L6] [ )] [ENTER]. ([L6] is the second function associated with the [6] key.) Now we know $\sum_{i=1}^{4}(c_i - \overline{x})(d_i - \overline{y}) = 18.8$.

```
NAMES OPS MATH          NAMES OPS MATH          sum(L6)
1:L1                    1:min(                              18.8
2:L2                    2:max(                  ■
3:L3                    3:mean(
4:L4                    4:median(
5:L5                    5:sum(
6:L6                    6:prod(
7↓COST                  7↓stdDev(
```

The slope is $\dfrac{18.8}{5} = 3.76$, thus the regression line is $y - 10.65 = 3.76(x - 2.5)$, which simplifies to $y = 3.76x + 1.25$.

Using the LINEAR REGRESSION feature is certainly quicker than this method, but they both achieve the same result.

Below the original table is filled in.

Variable	c_i	d_i	$c_i - \overline{x}$	$(c_i - \overline{x})^2$	$d_i - \overline{y}$	$(c_i - \overline{x})(d_i - \overline{y})$
Calculator location	L_1	L_2	$L_3 = $ $L_1 - \overline{x}$	$L_4 = L_3{}^2$	$L_5 = $ $L_2 - \overline{y}$	$L_6 = L_3 \bullet L_5$
	1	5.2	-1.5	2.25	-5.45	8.175
	2	8.9	-0.5	0.25	-1.75	0.875
	3	11.7	0.5	0.25	1.05	0.525
	4	16.8	1.5	2.25	6.15	9.225
	$\sum_{i=1}^{4} c_i = 10$	$\sum_{i=1}^{4} d_i = 42.6$		$\sum_{i=1}^{4}(c_i - \overline{x})^2 = 5$		$\sum_{i=1}^{4}(c_i - \overline{x})(d_i - \overline{y}) = 18.8$
	$\overline{x} = 2.5$	$\overline{y} = 10.65$				

The TI-86

Graphing Calculator

Chapter R
Functions, Graphs, and Models

GETTING STARTED

Press ON to turn on the calculator. (ON is the key at the bottom left-hand corner of the keypad.) You should see a blinking rectangle, or cursor, on the screen. If you do not see the cursor, try adjusting the display contrast. To do this, first press 2nd. (2nd is the yellow key in the upper left portion of the keypad.) Then press and hold ▲ to increase the contrast or ▼ to decrease the contrast. If the contrast needs to be adjusted further after the first adjustment, press 2nd and then hold ▲ or ▼ to increase or decrease the contrast, respectively.

Press 2nd [MODE] to display the MODE settings. ([MODE] is the second function associated with the MORE key.) Initially you should select the settings on the left side of the display.

To change a setting on the mode screen use ▲ or ▼ to move the cursor to the line of that setting. Then use ▶ or ◀ to move the blinking cursor to the desired setting and press ENTER. Press EXIT, CLEAR, or 2nd [QUIT] to leave the mode screen. ([QUIT]is the second function associated with the EXIT key.) In general, second functions are written in yellow above the keys on the keypad of the TI-86. The TI-86 also has an ALPHA key. It is the blue key directly below the 2nd key. When it is pressed, each of the blue alphabetic characters can be accessed on the calculator. Both upper and lower case letters are available on the TI-86. Pressing the 2nd key before the ALPHA key will produce lower case letters. Pressing EXIT, CLEAR, or 2nd [QUIT] will take you to the home screen where computations are performed.

It will be helpful to read the Quick Start section and Chapter 1 in your TI-86 Guidebook before proceeding.

USING A MENU

A menu is a list of options that appears when a key is pressed, thus, multiple options, and sometimes multiple menus, may be accessed by pressing one key. For example, the following screen appears when 2nd [MATH] is pressed. ([MATH] is the second function associated with the × key. This is the multiplication key.) We see several submenus at the bottom of the screen. The F1 - F5 keys at the top of the keypad are used to select options from this menu. The arrow to the right of "MISC" indicates that there are more choices. They can be seen by pressing MORE.

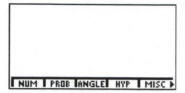 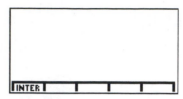

To choose the NUMBER submenu from the MATH menu press MORE again to return to the original menu and then press F1. When "NUM" is chosen, the original submenus move up on the screen and the items on the NUMBER submenu appear at the bottom of the screen.

 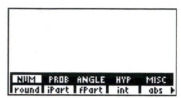

When two rows of options are displayed like this, the top row is accessed by pressing 2nd followed by one of the keys F1 - F5. These keystrokes access [M1] - [M5], the second operations, associated with the F1 - F5 keys. The options on the bottom row are accessed by pressing one of the keys F1 - F5. Absolute value, denoted "abs," is selected from the NUMBER submenu and copied to the home screen, for instance, by pressing F5.

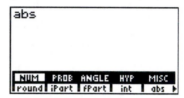

A menu can be removed from the home screen by pressing EXIT. If both a menu and a submenu are displayed, press EXIT once to remove the submenu and twice to remove both from the home screen.

SETTING THE VIEWING WINDOW

Section R.1, Technology Connection, page 7   (Page numbers refer to pages in the textbook.)

The viewing window is the portion of the coordinate plane that appears on the graphing calculator's screen. It is defined by the minimum and maximum values of x and y: xMin, xMax, yMin, and yMax. The notation [xMin, xMax, yMin, yMax] is used in the text to represent these window settings. For example, [-12, 12, -8, 8] denotes a window that displays the portion of the x-axis from -12 to 12 and the portion of the y-axis from -8 to 8. In addition, the distance between tick marks on the axes is defined by the settings xScl and yScl. In this manual, xScl and yScl will be assumed to be 1 unless noted otherwise. The setting xRes sets the pixel resolution. The default is xRes = 1. The window corresponding to the settings [-20, 30, -12, 20], xScl = 5, yScl = 2, xRes = 1, is shown below.

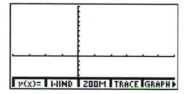

Press the GRAPH F2 to display the current window settings on your calculator. The standard viewing window is shown on the next page.

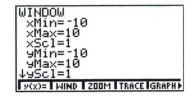
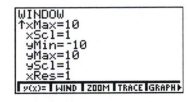

To change a setting, position the cursor to the right of the setting you wish to change and enter the new value. For example, to change from the standard settings to [-20, 30, -12, 20], xScl = 5, yScl = 2, use the ▲ and ▼ keys if necessary to position the cursor beside "xMin =" on the window screen then press (-) 2 0 ENTER 3 0 ENTER 5 ENTER (-) 1 2 ENTER 2 0 ENTER 2 ENTER. You must use the (-) key on the bottom row of the keypad rather than the − key on the right-hand side of the keypad to enter a negative number. (-) represents "the opposite of" or "the additive inverse of" a number, whereas − is the key used to subtract. The down-arrow, ▼, may be used instead of ENTER after typing each window setting. To see the window shown above, press the F5 key located on the top row of the keypad.

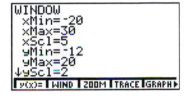
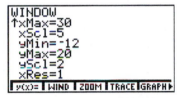
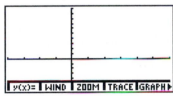

QUICK TIP: To return quickly to the standard viewing window [-10, 10, -10, 10], xScl = 1, yScl = 1, press GRAPH F3 F4.

GRAPHS

After entering an equation and setting a viewing window, you can view the graph of an equation.

Section R.1, Technology Connection, page 7   Graph $y = x^3 - 5x + 1$ using a graphing calculator.

Equations are entered on the y(x) =, or equation-editor, screen. Press GRAPH F1 to access this screen. If any plot (Plot 1, Plot 2, or Plot 3) is turned on (highlighted), turn it off by using the arrow keys to move the blinking cursor over the plot name and press ENTER. If there is currently an expression displayed for y1, clear it by positioning the cursor to the right of "y1 =" and pressing CLEAR. Remove all other expressions that appear in the equation-editor screen by pressing ▼ and CLEAR. Then use ▲ and ▼ to move the cursor to the right of "y1 =." Now press F1 ^ 3 − 5 F1 + 1 to enter the right-hand side of the equation on the equation-editor screen. The keystroke F1 can be replaced with x-VAR. When the equation-editor screen is displayed either keystroke will produce the variable x in an equation.

The standard viewing window, [-10, 10, -10, 10], is a good choice for this graph. Either press [2nd] [F2] enter these dimensions, [-10, 10, -10, 10], on the window screen and then press [F5] or simply press [GRAPH] [F3] [F4] to select the standard viewing window and see the graph.

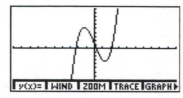

You can edit your entry if necessary. If, for instance, you pressed [6] instead of [5], when you put the equation into y1, you can use the [◄] key to move the cursor to 6 and then press [5] to overwrite it. If you forgot to type the plus sign, move the cursor to the 1, and then press [2nd] [INS] [+] to insert the plus sign before the 1. ([INS] is the second function associated with the [DEL] key.) You can continue to insert symbols immediately after the first insertion without pressing [2nd] [INS] again. If you typed 52 instead of 5, move the cursor to the 2 and press [DEL] to delete the 2.

An equation must be solved for y before it can be graphed on the TI-86.

Section R.1, Technology Connection, page 8; Example 2, page 5   To graph $3x + 5y = 10$, first solve for y, obtaining $y = \dfrac{-3x + 10}{5}$. Then press [GRAPH] [F1] and clear any expressions that currently appear. Position the cursor to the right of "y1 =." Press [(] [(-)] [3] [F1] [+] [1] [0] [)] [÷] [5] to enter the right-hand side of the equation. Note that without the parentheses the expression $-3x + \dfrac{10}{5}$ would have been entered. As before, [F1] can be replaced with [x-VAR] to produce the variable x in the equation.

Press [GRAPH] [F3] [F4] to see the graph displayed using the standard viewing window. You may change the viewing window as desired to reveal more or less of the graph. The graph is shown here using the standard viewing window.

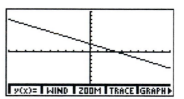

Section R.1, Technology Connection, page 8; Example 4, page 6   To graph $x = y^2$, first solve the equation for y, obtaining $y = \pm\sqrt{x}$. To see the entire graph of $x = y^2$ we must graph y1 $= \sqrt{x}$ and y2 $= -\sqrt{x}$ on the same screen. Press [GRAPH] [F1] and clear any expressions that currently appear. With the cursor to the right of "y1 =" press [2nd] [√] [F1] or [2nd] [√] [x-VAR]. ([√] is the second function associated with the [x²] key.)

Now use [▼] or press [ENTER] to move the cursor to the right of "y2 =." Here are three ways to enter y2 $= -\sqrt{x}$.

The first method is to enter the expression $-\sqrt{x}$ directly by pressing [(-)] [2nd] [√] [F1] or [(-)] [2nd] [√] [x-VAR].

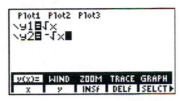

The second method of entering $y2 = -\sqrt{x}$ is based on the fact that $-\sqrt{x}$ is the additive inverse of $\sqrt{x}$. In other words, $y2 = -y1$. To enter this, move the cursor to the right of "y2 =" and press [(-)] [F2] [1].

The third method of entering $y2 = -\sqrt{x}$ involves using the RECALL feature. Move the cursor to the right of "y2 =" and press [(-)] [2nd] [RCL]. ([RCL] is the second function associated with the [STO▸] key.) Notice that the flashing cursor, [A], is flashing. This indicates that the calculator is in ALPHABETIC mode. To turn off ALPHABETIC mode, press the [ALPHA] key. Press [F2] [1] [ENTER] and $\sqrt{x}$ is pasted on the screen. This method is particularly useful if the expression to be copied is lengthy.

Once both equations are on the equation-editor screen, select a viewing window by pressing [GRAPH] [F2]. The window shown here is [-2, 10, -5, 5]. Press [F5] to display the graph.

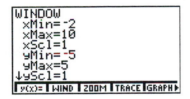   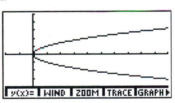

The top half is the graph of y1, the bottom half is the graph of y2, and together they yield the graph of $x = y^2$.

THE TABLE FEATURE

For an equation entered in the equation-editor screen, a TABLE of x- and y-values can be displayed.

Section R.2, Technology Connection, page 17 Create a table of ordered pairs for the function $f(x) = x^3 - 5x + 1$.

Enter the function as $y1 = x^3 - 5x + 1$ as described on page 83 of this manual. Once the equation is entered, press TABLE F2 to display the table set-up screen. A starting value for x can be chosen along with an increment for the x-value. To select a starting value of 0 and an increment of 1, press 0 and ENTER or ▼ and then press 1. The "Indpnt:" setting should be "Auto." If it is not, use the ▼ to position the blinking cursor over "Auto" and then press ENTER. To display the table, press F1.

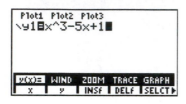  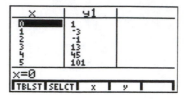

Use the ▲ and ▼ keys to scroll through the table. For example, by using ▼ to scroll down we can see that $y1 = 1277$ when $x = 11$. Using ▲ to scroll up, observe that $y1 = -11$ when $x = -3$.

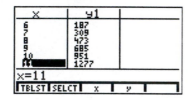   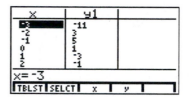

GRAPHS and FUNCTION VALUES

Section R.2, Technology Connection, page 19   There are several ways to evaluate a function using a graphing calculator. Three of them are described here. Given the function $f(x) = 2x^2 + x$, we will find $f(-2)$.

First press GRAPH F1 and enter the function as $y1 = 2x^2 + x$. Now we will find $f(-2)$ using the TABLE feature. Press TABLE F2. Move the cursor to "Indpnt:" and highlight "Ask." Press ENTER. When "Ask" is highlighted, you supply the x-values and the graphing calculator returns the corresponding y-values. The settings for "TblStart" and "$\triangle$ Tbl" are irrelevant in this mode. Press F1 and find $f(-2)$ by pressing (-) 2 ENTER. We see that $y1 = 6$ when $x = -2$, so $f(-2) = 6$.

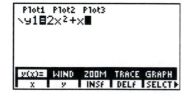  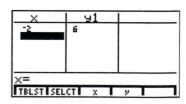

We can also use the EVALUATE feature from the GRAPH menu to find $f(-2)$. To do this, graph $f(x) = 2x^2 + x$ in a window that includes the x-value, -2. Press GRAPH F3 F4 and we see the graph displayed on the standard viewing window. Press GRAPH MORE MORE F1 to select "EVAL." Press (-) 2 ENTER and "x = -2" and "y = 6" appear at the bottom of the screen. Again we see that $f(-2) = 6$.

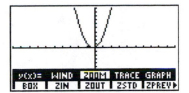
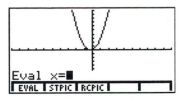
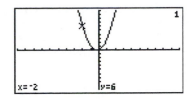

The third method for finding $f(-2)$ uses function notation on the home screen of the calculator. With

$y1 = 2x^2 + x$ entered on the equation-editor screen, go to the home screen by pressing [2nd] [QUIT]. ([QUIT] is the

second function associated with the [EXIT] key.) Now enter $y1(-2)$ by pressing [2nd] [alpha] Y. (Y is the alphabetic

character associated with the [0] key.) The letter Y will appear on the screen in upper case if the [ALPHA] Y is pressed

or in lower case if [2nd] [ALPHA] Y is pressed. Press [1] [(] [(-)] [2] [)] [ENTER]. Again we see that $f(-2) = 6$.

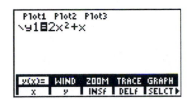
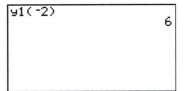

THE TRACE FEATURE

The TRACE feature can be used to display the coordinates of points on a graph.

Section R.2, Technology Connection, page 19   Use the function $f(x) = 2x^2 + x$ graphed in the standard viewing

window. Press [GRAPH] [F1] and put the function on the equation-editor screen. Then press [GRAPH] [F3] [F4].

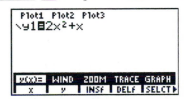
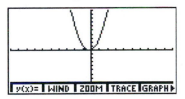

Press [GRAPH] [F4] and a blinking cursor appears on the graph at the x-value that is the x-coordinate of the

midpoint of the x-axis. The y-coordinate associated with this x-coordinate is also displayed.

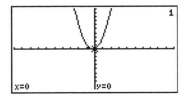

Press [▶] and another ordered pair will be displayed at the bottom of the screen.

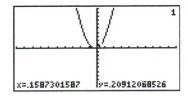

Press [3] [ENTER] to see that when $x = 3$ the associated y-coordinate is 21. Even though 21 is larger than

maximum y-value in the window, it is still displayed at the bottom of the screen.

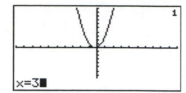   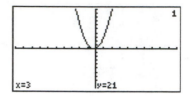

If you enter an *x*-value that is not between the minimum and maximum window values for *x*, an error will result. Try it! Press [2] [0] [ENTER] and an error results. Press [F5] to select "QUIT."

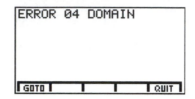

GRAPHING FUNCTIONS DEFINED PIECEWISE; DOT MODE

Section R.2, Technology Connection, page 23; Example 9, page 22   Graph: $f(x) = \begin{cases} 4 & for \ \ x \le 0 \\ 3 - x^2 & for \ \ 0 < x \le 2 \\ 2x - 6 & for \ \ x > 2 \end{cases}$.

We will enter the function using inequality symbols from the TEST menu. Press [GRAPH] [F1] to go to the equation-editor screen. Clear any entries that are present. Position the cursor to the right of "y1=" and press [(] [4] [)] [(] [x-VAR] [2nd] [TEST] [F4] [0] [)] [+] [(] [3] [-] [x-VAR] [x²] [)] [(] [(] [0] [F2] [x-VAR] [2nd] [BASE] [F4] [F1] [x-VAR] [2nd] [TEST] [F2] [2] [)] [+] [(] [2] [x-VAR] [-] [6] [)] [(] [x-VAR] [F3] [2] [)]. ([TEST] is the second function associated with the [2] key and [BASE] is the second function associated with the [1] key.) The keystrokes [2nd] [TEST] [F4] open the TEST menu and select "≤" from the menu. The keystrokes [2nd] [TEST] [F2] open the TEST menu and select " <" from the menu. With the TEST menu displayed, we select " >" by pressing [F3]. The keystrokes [2nd] [BASE] [F4] [F1] select "and" from the BOOLEAN submenu ("BOOL") of the BASE menu.

When graphing a piece-wise defined function, put the calculator in DRAWDOT mode. If this is not done, a vertical line that is not part of the graph could appear. DRAWDOT mode can be selected in two ways. One is to press [GRAPH] [MORE] [F3]. On this screen we can format the graph. Move the cursor over "DrawDot", and press [ENTER]. To leave this screen press [2nd] [QUIT]. DRAWDOT mode can also be selected on the equation-editor screen by pressing [GRAPH] [F1] [MORE]. Now press [F3] six times to change the style until the dotted graph style icon appears to the left of "y1 =."

Choose and enter window dimensions by pressing [GRAPH] [F2]. Change the window to the one shown on the top of the next page on the left. Then press [F5] to display the graph of the function using these window settings.

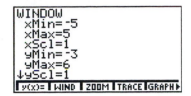 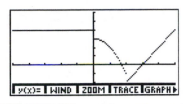

To return to CONNECTED, or DRAWLINE, mode press GRAPH MORE F3, highlight "DrawLine," and press ENTER. When the equation in y1 is cleared, the graph style will automatically default back to CONNECTED mode.

SQUARING THE VIEWING WINDOW

Section R.4, Technology Connection, page 40 In the standard viewing window, the distance between tick marks on the *y*-axis is about $\frac{3}{5}$ the distance between tick marks on the *x*-axis. It is often desirable to choose window dimensions for which these distances are the same, creating a "square" window. Any window in which the ratio of the length of the *y*-axis to the length of the *x*-axis is $\frac{3}{5}$ will produce this effect. This can be accomplished by selecting dimensions for which yMax – yMin = $\frac{3}{5}$ (xMax – xMin).

The standard viewing window is shown on the left below and the square window [-10, 10, -6, 6] is shown on the right. Observe that the distance between tick marks appears to be the same on both axes in the square window.

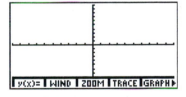 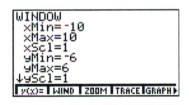 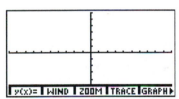

The window can also be squared by pressing GRAPH F3 MORE F2 to select "ZSQR" (zoom square) from the GRAPH ZOOM menu. Starting with the standard viewing window and pressing GRAPH F3 MORE F2 produces the dimensions and the window shown below.

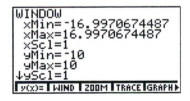 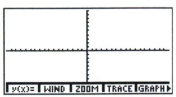

THE INTERSECT FEATURE

Section R.4, Example 10, page 49 *Business: Profit-and-Loss Analysis* When a business sells an item, it receives the price paid by the consumer. This is normally greater than the cost to the business of producing the item.

(a)   The total revenue that a business receives is the product of the number of items sold and the price paid per item. Thus, if Raggs, Ltd. sells x suits at $80 per suit, the total revenue, $R(x)$, is given by $R(x) = 80x$. The cost equation, $C(x) = 20x + 10,000$, is found in Example 9 on page 48 in the text book. Graph both $R(x)$ and $C(x)$ on the same set of axes.

The window must be modified to fit the problem. Certainly it makes no sense to have a negative number of products; therefore, since the x-axis is the item axis, the minimum value should be 0. The y-axis is the money axis and can certainly be negative, but for this first graph is not. Below are the window settings and the graphs of the revenue and cost equations.

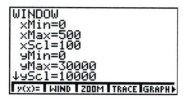   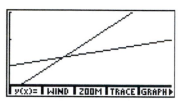

(b)   The total profit a business makes is the money that is left after all costs have been subtracted from the total revenue. If $P(x)$ represents the profits when x items are produced and sold, we have $P(x) = R(x) - C(x)$. Determine $P(x)$ and draw its graph using the same set of axes as those used for the graph in part (a).

The revenue equation is in y1 and the cost equation is in y2, therefore $P(x) = $ y1 $-$ y2. Position the cursor immediately to the right of "y3 =." Press [F2] [1] [−] [F2] [2]. To allow for the profit to be negative, the window has been modified and all three graphs are displayed.

   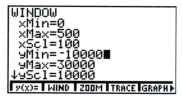   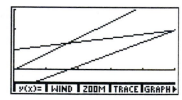

(c)   The company will break even at that value of x for which $P(x) = 0$ (that is, no profit and no loss). This is the point at which $R(x) = C(x)$. Find the break-even point.

Use the INTERSECT feature of the GRAPH MATH menu to find the break-even point. With the graph displayed, press [GRAPH] [MORE] [F1] [MORE] [F3] to choose "ISECT." The calculator will pose three questions.

The query "First Curve?" appears at the bottom of the screen. The blinking cursor is positioned on the graph of y1 as in the screen on the top of the next page on the left. Notice the upper right corner of this screen displays a 1, indicating that the cursor is on y1. Press [ENTER] to indicate that this is the first curve involved in the intersection.

Next the query "Second Curve?" appears at the bottom of the screen. The blinking cursor is now positioned on the graph of y2 as in the screen below on the right. Notice the upper right corner of this screen displays a 2, indicating that the cursor is on y2. Press [ENTER] to indicate that this is the second curve. We identify the curves for the calculator since we have three graphs on the screen.

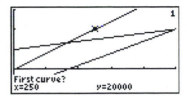
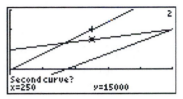

After we identify the second curve, the query "Guess?" appears at the bottom of the screen. Since there is only one point of intersection, press [ENTER] and the point of intersection is displayed. This is the break-even point.

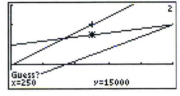
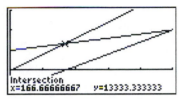

After finding the break-even point, press [x-VAR] [ENTER] [GRAPH] [F4]. Then press [▼] until and you see the screen below. Notice the number 3 in the upper right corner, indicating the cursor is on y3. This screen shows the value of the profit function when the x-coordinate of the break-even, or equilibrium, point is substituted into it. The calculator has written the y value in scientific notation. This number is $2 \times 10^{-8} = 0.00000002$. This number is really 0, so the profit is $0 at the break-even point.

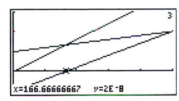

The company breaks even when 167 items are produced and sold.

THE MAXIMUM FEATURE

The MAXIMUM feature from the GRAPH MATH menu can be used to find the vertex of a parabola on the graph screen.

Section R.5, Example 2, page 57 Graph $f(x) = -2x^2 + 10x - 7$.

Graph the function as $y1 = -2x^2 + 10x - 7$ on the standard viewing window. The screen at the top of the next page on the left should appear. Press [GRAPH] [MORE] [F1] to go to the GRAPH MATH menu. To put the graph on graph paper, you need to find a few points. The most important point is the vertex. Since this parabola is concave down, the vertex is its maximum point. Press [F5] to choose "FMAX." In the bottom left corner of the screen "Left Bound?" appears. Move the cursor using the left- and/or right-arrow keys until the cursor is in a location on the curve that is to the left of the vertex then press [ENTER]. When you press [ENTER] you will see a marker appear on the screen to mark the left boundary and "Right Bound?" will appear on the bottom of the screen.

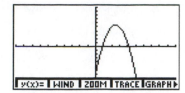 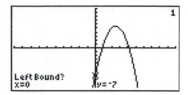 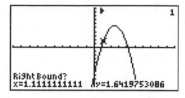

Move the cursor to the right of the vertex and press ENTER. Another marker appears, this one marks the right boundary, and "Guess?" appears on the bottom of the screen. Move the cursor near the vertex of the parabola and press ENTER. The vertex is displayed.

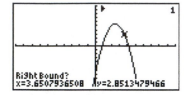 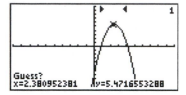 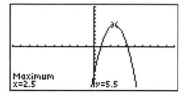

Sometimes the calculator does not give the exact answer. It is worth noting that the *x*- and *y*-values are temporarily stored in memory. Go to the home screen by pressing 2nd [QUIT]. The *x*-coordinate is 2.5. Press x-VAR and turn it into a fraction by pressing 2nd [MATH] F5 MORE F1 to choose "Frac" and press ENTER. To find the *y*-value press 2nd [alpha] Y 1. (Y is the alphabetic character associated with the 0 key. Pressing 2nd [alpha] produces lower case letters.) Press 2nd [MATH] F5 MORE F1 to choose "Frac" and press ENTER to turn it into a fraction.

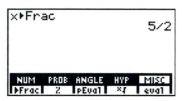 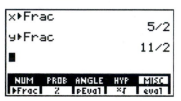

To complete the graph you need more points than just the vertex. Use the table to find more points. Press GRAPH F5 to see the graph. Press TABLE F2 to display the table set-up screen. A starting value of *x* can be chosen along with an increment for the *x*-value. To select a starting value of 0 and an increment of 1, press 0 and ENTER or ▼ and then press 1. Set "Indpnt:" to "Auto." If it is not set to "Auto," use the ▼ to position the blinking cursor over "Auto" and then press ENTER. To display the table, press F1.

Notice the vertex is not on the table since the increment is 1, but you can tell that the *x*-coordinate of the vertex is half way between the *x*-values 2 and 3.

THE POLYNOMIAL ROOT-FINDER

The following example will be solved using the POLYNOMIAL ROOT-FINDER and will be solved graphically using the INTERSECT feature on page 94 of this manual. It is solved algebraically in the text book.

Section R.5, Example 3, page 58   Solve: $3x^2 - 4x = 2$.

First write the equation in standard form $ax^2 + bx + c = 0$, and then determine a, b, and c:

$3x^2 - 4x - 2 = 0$, so $a = 3$, $b = -4$, and $c = -2$.

Press 2nd [POLY] to access the POLYNOMIAL ROOT-FINDER. ([POLY] is the second function associated with the PRGM key.) The order of this polynomial is 2. Press 2 ENTER. In the calculator a_2 is the coefficient of x^2, a_1 is the coefficient of x, and a_0 is the constant. Enter the coefficients as in the last screen below.

   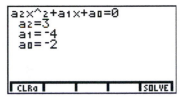

Notice the choices along the bottom of the calculator screen. "SOLVE" is above the F5 key. Press F5 and the equation's solutions appear on the screen. The decimal approximations of the answers are "x_1" and "x_2" on the screen below.

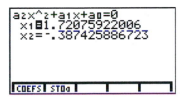

To save these answers move the cursor over the "=" sign next to the root value you want to store and press STO►. The ALPHA-lock is on. The cursor looks like a flashing ▣. Press ROOT. (R, O, and T are the alphabetic characters associated with the 5, ×, and − keys, respectively.) To remove the ALPHA-lock, press ALPHA.

   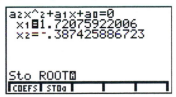   

Press 1 ENTER. Press 2nd [QUIT] to return to the home screen. To see if the answer was saved, press 2nd [RCL]. ([RCL] is the second function associated with the STO► key.) Notice the ALPHA-lock cursor is displayed again. Press ROOT. (R, O, and T are the alphabetic characters associated with the 5, ×, and − keys, respectively.) To remove the ALPHA-lock, press ALPHA. Then press 1.

   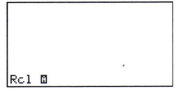   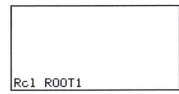

To see the answer displayed on the home screen, press ENTER.

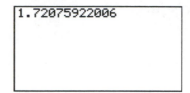

THE INTERSECT FEATURE

We can use the INTERSECT feature from the GRAPH MATH menu to solve equations on the graph screen.

Section R.5, Example 3, page 58   Solve: $3x^2 - 4x = 2$.

Now we will solve this problem graphically. If we consider the left and right sides of the equation as two separate equations, the solutions to the equation will be the *x*-coordinates of the points of intersection.

Press $\boxed{\text{GRAPH}}$ $\boxed{\text{F1}}$ clear any existing entries and enter $y1 = 3x^2 - 4x$ and y2 = 2. Press $\boxed{\text{GRAPH}}$ $\boxed{\text{F3}}$ $\boxed{\text{F4}}$ to display the graph on a standard viewing window.

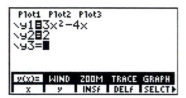   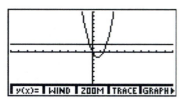

Use the INTERSECT feature from the GRAPH MATH menu to find the points of intersection. Press $\boxed{\text{GRAPH}}$ $\boxed{\text{MORE}}$ $\boxed{\text{F1}}$ $\boxed{\text{MORE}}$ $\boxed{\text{F3}}$ to choose "ISECT." The calculator will pose three questions.

The query "First Curve?" appears at the bottom of the screen. The blinking cursor is positioned on the graph of y1 as in the screen below on the left. Notice the upper right corner of this screen displays 1. This indicates that the cursor is on y1. Press $\boxed{\text{ENTER}}$ to indicate that this is the first curve involved in the intersection.

Next the query "Second Curve?" appears at the bottom of the screen. The blinking cursor is now positioned on the graph of y2 as in the screen below on the right. Notice the upper right corner of this screen displays a 2, indicating that the cursor is now on y2. Press $\boxed{\text{ENTER}}$ to indicate that this is the second curve.

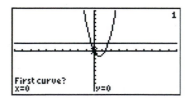   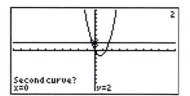

After we identify the second curve, the query "Guess?" appears at the bottom of the screen. Since the cursor is near one of the points of intersection, press $\boxed{\text{ENTER}}$ and the point of intersection nearest the guess is displayed.

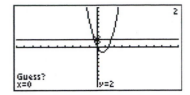 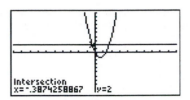

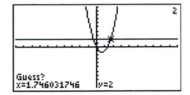 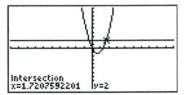

To find the second point of intersection, repeat the process. Make sure you move the cursor near the second point of intersection before you press ENTER.

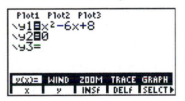 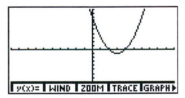

The two answers, rounded to three decimal places, are $x \approx -0.387$ or $x \approx 1.721$.

THE INTERSECT FEATURE

Another example of how the INTERSECT feature from the GRAPH MATH menu can be used to solve equations.

Section R.5, Technology Connection, page 59 Given, $f(x) = x^2 - 6x + 8$, find the solutions of $x^2 - 6x + 8 = 0$.

If we consider the left and right sides of the equation as two separate equations, the solutions to the equation will be the x-coordinates of the points of intersection.

Press GRAPH F1 clear any existing entries and enter $y1 = x^2 - 6x + 8$ and $y2 = 0$. Press GRAPH F3 F4 to display the graph on a standard viewing window. Remember that $y2 = 0$ is the graph of the x-axis, so you will not see another line since the axis is already on the screen.

Use the INTERSECT feature from the GRAPH MATH menu to find the points of intersection. Press GRAPH MORE F1 MORE F3 to choose "ISECT." The calculator will pose three questions.

The query "First Curve?" appears at the bottom of the screen. The blinking cursor is positioned on the graph of y1 as in the screen on the next page on the left. Notice the upper right corner of this screen displays a 1, indicating that the cursor is positioned on y1. Press ENTER to indicate that this is the first curve involved in the intersection.

Next the query "Second Curve?" appears at the bottom of the screen. The blinking cursor is now positioned on the graph of y2 as in the screen on the next page on the right. Notice the upper right corner of this screen displays a 2, indicating that the cursor is on y2. Press ENTER to indicate that this is the second curve. Notice that the cursor is on the x-axis. Again, that is because $y2 = 0$ is the graph of the x-axis.

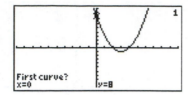
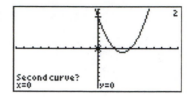

After we identify the second curve, the query "Guess?" appears at the bottom of the screen. Since the cursor is near one of the points of intersection, press ENTER and the point of intersection nearest the guess is displayed.

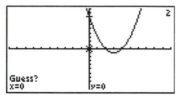
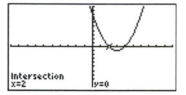

To find the second point of intersection, repeat the process. Make sure you move the cursor near the second point of intersection before you press ENTER.

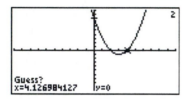
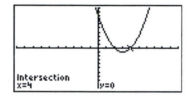

The two answers are $x = 4$ or $x = 2$.

THE INTERSECT FEATURE

Another example of how the INTERSECT feature from the GRAPH MATH menu can be used to solve equations.

Section R.5, Technology Connection, page 61 Solve the equation $x^3 = 3x + 1$.

If we consider the left and right sides of the equation as two separate equations, the solutions to the equation will be the x-coordinates of the points of intersection.

On the equation-editor screen, clear any existing entries and enter $y1 = x^3$ and $y2 = 3x + 1$. Now graph these equations in an appropriate window. One good choice is [-3, 3, -10, 10].

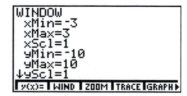
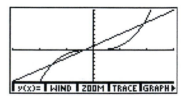

Use the INTERSECT feature from the GRAPH MATH menu to find the points of intersection. Press GRAPH MORE F1 MORE F3 to choose "ISECT." The calculator will pose three questions.

The query "First Curve?" appears at the bottom of the screen. The blinking cursor is positioned on the graph of y1 as in the screen on the top of the next page on the left. Notice the upper right corner of this screen displays a 1, indicating that the cursor is on y1. Press [ENTER] to indicate that this is the first curve involved in the intersection.

Next the query "Second Curve?" appears at the bottom of the screen. The blinking cursor is now positioned on the graph of y2 as in the screen below on the right. Notice the upper right corner of this screen displays a 2, indicating that the cursor is on y2. Press [ENTER] to indicate that this is the second curve.

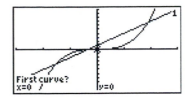
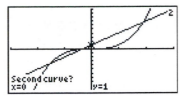

After we identify the second curve, the query "Guess?" appears at the bottom of the screen. Move the cursor near the leftmost point of intersection. Press [ENTER]. The point of intersection nearest the guess is displayed.

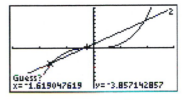
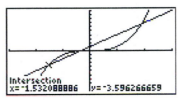

To find the other two points of intersection repeat the process for each point. Make sure you move the cursor near the point of intersection you are trying to find before you press [ENTER].

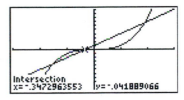
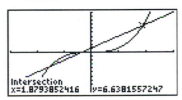

The three answers, rounded to two decimal places, are $x \approx -1.53$, $x \approx -0.35$ or $x \approx 1.88$.

THE ROOT FEATURE

Here is an example of how an equation can be solved on the graph screen using the ROOT feature from the GRAPH MATH menu. The equation to be solved must first be expressed in the form $f(x) = 0$. Many students prefer to use the INTERSECT feature, as described on page 94, 95, and 96 of this manual, instead of the ROOT feature.

Section R.5, Technology Connection, page 61 Solve the equation $x^3 = 3x + 1$.

To use the ROOT feature, we must first have the equation in the form $f(x) = 0$. To do this, subtract $3x + 1$ on both sides of the equation to obtain an equivalent equation with 0 on one side. We have $x^3 - 3x - 1 = 0$. The solutions of the equation $x^3 = 3x + 1$ are the values of x for which the function $f(x) = x^3 - 3x - 1$ is equal to zero. We can use the ROOT feature to find these values, or roots. In some texts, you will see the roots referred to as the zeros. These terms are interchangeable.

On the equation-editor screen, clear any existing entries and then enter $y1 = x^3 - 3x - 1$. Now graph the function in a viewing window that shows the *x*-intercepts clearly. One good choice is [-3, 3, -5, 8]. We see that the function has three roots or zeros. They appear to be about -1.5, -0.5, and 2.

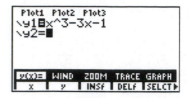   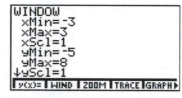   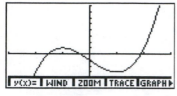

We will find the root near -1.5 first. Press MORE F1 F1 to select "ROOT" from the GRAPH MATH menu. We are prompted to select a left bound. This means that we must choose an *x*-value that is to the left of -1.5 on the *x*-axis. This can be done by using the left- and right-arrow keys to move the cursor to a point on the curve to the left of -1.5 or by keying in a value less than -1.5.

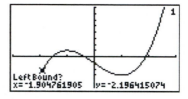

Once this is done press ENTER. Now we are prompted to select a right bound that is to the right of -1.5 on the *x*-axis. Again, this can be done by using the arrow keys to move the cursor to a point on the curve to the right of −1.5 or by keying in a value greater than -1.5.

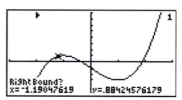

Press ENTER again. Finally we are prompted to make a guess as to the value of the root. Move the cursor to a point near the root or key in a value.

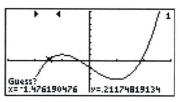

Press ENTER a third time. We see that $y = 0$ when $x \approx -1.53$, so -1.53 is a root of the function.

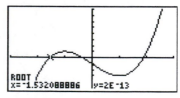

Select "ROOT" from the GRAPH MATH menu a second time to find the root near -0.5 and a third time to find the root near 2.

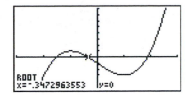 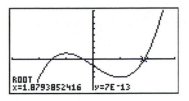

We see that the other two roots or zeros are approximately -0.35 and 1.88.

ABSOLUTE-VALUE FUNCTIONS

We can use the absolute-value option to perform computations involving absolute value and to graph absolute-value functions.

Section R.5, Example 8, page 65 Graph $f(x) = |x|$.

The absolute-value function is found in the MATH NUMBER menu. To graph $f(x) = |x|$ press [GRAPH] [F1] to go to the equation-editor screen and then clear any existing entries. Position the cursor to the right of "y1 =" and press [2nd] [MATH] [F1] [F5] to copy "abs" to the equation-editor screen. ([MATH] is the second function associated with the [×] key.) Then press [x-VAR]. Choose an appropriate viewing window and graph the function. To use the standard window, press [GRAPH] [F3] [F4].

 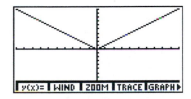

GRAPHING RADICAL FUNCTIONS

There are various ways to enter radical expressions on a graphing calculator.

Section R.5, Powers with Rational Exponents, page 66 We discussed entering an expression containing a square root on page 84 of this manual. If the radicand has more than one term, the entire radicand must be enclosed in parentheses. To enter $y1 = \sqrt{x+2}$, for example, position the cursor immediately to the right of "y1 =" on the equation-editor screen and press [2nd] [√] [(] [x-VAR] [+] [2] [)].

Higher order radical expressions can be entered using "$\sqrt[x]{}$" from the MATH MISCELLANEOUS menu. We must enclose the radicand in parenthesis if it contains more than one term. To enter $y1 = \sqrt[3]{x-2}$, position the cursor immediately to the right of "y1 =" on the equation-editor screen. Then press [3] to indicate that we are entering a cube root. Next press [2nd] [MATH] [F5] [MORE] [F4] to select "$\sqrt[x]{}$." Finally press [(] [x-VAR] [−] [2] [)] to enter the radicand.

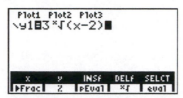

Press [GRAPH] [F2] and change your settings to the ones below. Then press [F5].

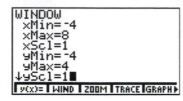   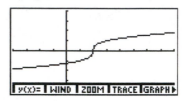

Remember that $\sqrt[3]{x-2} = (x-2)^{\frac{1}{3}}$. Return to the equation-editor screen by pressing [GRAPH] [F1] and with the cursor to the right of "y2 =" press [(] [x-VAR] [−] [2] [)] [^] [(] [(] [1] [÷] [3] [)]. Before graphing this function, press [GRAPH] [F1] [MORE] with the cursor to the right of the "y2 =" and press [F3] four times, until you see "⬦." This icon will be referred to as the path icon in this manual. Now press [GRAPH] [F5] to see what looks like a ball moving along the curve that was already graphed. This shows that the two ways of writing cube root are equivalent.

   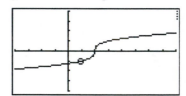

We could also compare the tables and see that the functions are the same.

To enter $y1 = \sqrt[4]{x-1}$, position the cursor to the right of "y1 =." Then press [4] to indicate that we are entering a fourth root. Next press [2nd] [MATH] [F5] [MORE] [F4] to select "$\sqrt[x]{}$ " from the MATH MISCELLANEOUS menu and finally press [(] [x-VAR] [−] [1] [)] to enter the radicand. The fourth root can also be written using a fractional exponent. To enter the alternate form of the function, position the cursor to the right of "y2 =" and press [(] [x-VAR] [−] [1] [)] [^] [(] [(] [1] [÷] [4] [)]. Prove to yourself they are the same by changing the graph style icon on one of them to path and graphing them or by examining the table. Here we examine the table. Press [TABLE] [F2] and set your table as in the middle screen below. Then press [F1] to see the table displayed. Notice the entries are the same. The ERROR message is because when taking an even root, the radicand's value can not be less than 0. This function does not exist when $x < 1$.

LINEAR REGRESSION

We can use the LINEAR REGRESSION feature from the STATISTICAL CALCULATIONS menu to fit a linear equation to a set of data points.

Section R.6, Technology Connection, page 78; Example 2, page 77 The following table shows the average annual pay for a U. S. production worker.

Years, x, since 1996	1	2	3	4	5	6	7
Percentage increase since 1996, P	1.9	7.4	11.7	19.5	28.2	29.7	31.3

(a)   Make a scatter plot of the data and determine whether the data seem to fit a linear function.

We will enter the data as ordered pairs on the statistical list-editor screen. To go to this screen and clear any existing lists first press [2nd] [STAT] [F2]. ([STAT] is the second function associated with the [+] key.) Then use the arrow keys to move up to highlight "xStat" and press [CLEAR] [ENTER]. Do the same for "yStat."

Once the lists are cleared, we can enter the data points. We will enter the number of years since 1996 in xStat and the percentage increase in yStat. Position the cursor just below "xStat." Press [1] [ENTER]. Continue typing the x-values 2 through 7, each followed by [ENTER]. The entries can be followed by [▼] rather than [ENTER] if desired. Press [▶] to move the cursor just below "yStat." Type the percentages 1.9, 7.4, and so on in succession, each followed by [ENTER] or [▼]. Note that the coordinates of each point must be in the same position in each list. The lists are too long to be seen on one screen. Use the up- and down-arrow keys to scroll through the lists.

To plot the data points, we turn on the STATISTICAL PLOT feature. To access the statistical plot screen, press [2nd] [STAT] [F3]. ([STAT] is the second function associated with the [+] key.)

Press [F1] to select "Plot 1." The cursor should be positioned over "On" and it should be flashing. Press [ENTER] to turn on Plot 1.

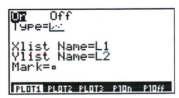

Press [▼] to move the cursor to the right of "Type=" and look at the choices at the bottom of the screen.

We are going to do a scatter plot. Select "SCAT" by pressing [F1]. The icon representing a scatter plot is displayed to the right of "Type =."

Press [▾] to move the cursor to the right of "Xlist Name =" and look at the choices at the bottom of the screen.

Press [F1] to choose "xStat." The *x*-coordinates of the data points are in xStat. Press [▾] to move the cursor to the right of "Ylist Name =" and press [F2] to choose "yStat."

Press [▾] to move the cursor to the right of "Mark=" and press [F1] to choose the mark that looks like a box.

The plot can also be turned on from the equation-editor screen. Press [GRAPH] [F1] to go to this screen. Before viewing the plot, any existing entries on the equation-editor screen should be cleared. Then assuming that Plot 1 has not yet been turned on and that the desired settings are currently entered for Plot 1 on the statistical plot screen, you can position the cursor over "Plot 1" and press [ENTER]. "Plot 1" will be highlighted. The easiest way to choose a viewing window is to use the ZOOM DATA feature. Press [GRAPH] [F3] [MORE] [F5] to select "ZDATA." The ZOOM DATA feature redefines the viewing window to display all statistical data points. Press [CLEAR] to remove the menu at the bottom of the screen.

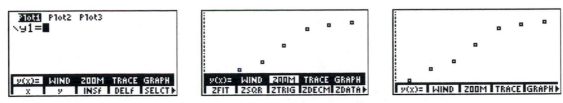

With all the data points in the viewing window it is easy to press GRAPH F2 and modify the window to suit your needs. Below the window has been changed and the new graph is displayed by pressing F5 CLEAR.

From the scatter plot, the data appears to be approximately linear.

(b)   Find a linear function that (approximately) fits the data.

The calculator's LINEAR REGRESSION feature can be used to fit a linear equation to the data. From the home screen and with the data points in the lists, press 2nd [STAT] F1 to choose "CALC." Press F3 to select "LinR."

Now press 2nd [LIST] F3 to see the list of all the named lists in the calculator. ([LIST] is the second function associated with the − key.)

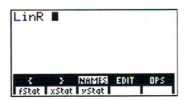

This list will vary from calculator to calculator, depending on who used it and what they named the lists. We need to choose "xStat" and "yStat." You may need to press MORE if there are several named lists in the calculator you are using. Press the appropriate F keys to select "xStat" and "yStat." Remember to use a comma to separate them. (Even though the calculator was programmed to assume the *x*- and *y*-coordinates of data points are in xStat and yStat, respectively, it is good practice to specify the lists being used.) Before the regression equation is found, it is possible to select a *y*-variable to which it will be stored on the equation-editor screen. To do this, press , 2nd [alpha] Y. (Y is the alphabetic character associated with the 0 key.) As before, it is lower case because the 2nd key was pressed first. Now press 1. Finally to display the coefficients of the regression equation press ENTER.

 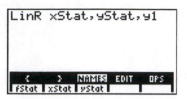 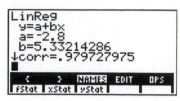

To remove the menus from the bottom of the screen, press [EXIT] [EXIT]. Note that we also see the coefficient of correlation, denoted "corr." This number indicates how well the regression line fits the data. The closer the absolute value of this number is to 1, the better the fit. We also see that "n = 7." This indicates that 7 data points were used to find the regression equation.

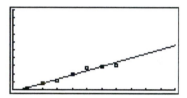

Rounding the coefficient of *x* to the nearest hundredth, we have $y = 5.33x - 2.8$.

Press [GRAPH] [F5] [CLEAR] to see the line displayed with the data points and to remove the menu at the bottom of the screen.

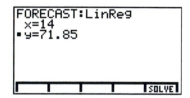

(c)   Use the model to predict the percentage by which 2010 wages will exceed 1996 wages.

To predict the percentage by which 2010 wages will exceed 1996 wages, evaluate the regression equation for *x* = 14. (2010 is 14 years after 1996.) Use any of the methods for evaluating a function presented earlier in this chapter. We will use the FORECAST feature from the STATISTICAL menu. Press [2nd] [STAT] [MORE] [F1]. The forecast screen will be displayed. With the cursor beside "x =," press [1] [4] [ENTER]. The cursor moves beside "y =." Press [F5] to select "SOLVE."

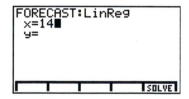 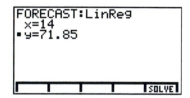

When $x = 14$, $y \approx 71.85$, so we predict 2010 wages will exceed 1996 wages by approximately 71.85%

POLYNOMIAL REGRESSION; Quadratic and Cubic

The TI-86 graphing calculator has the capability to use the REGRESSION feature to fit quadratic, cubic, and other functions to data.

Section R.6, Technology Connection, page 81 The chart on the following page relates the number of live births to women of a particular age.

Age, x	Average Number of Live Births Per 1000 Women
16	34
18.5	86.5
22	111.1
27	113.9
32	84.5
37	35.4
42	6.8

(a)   Fit a quadratic function to the data using the REGRESSION feature on the calculator. Then make a scatter plot of the data and graph the quadratic function with the scatter plot.

Clear any existing entries on the statistical list-editor screen. Enter the data into lists with the ages in xStat and the average number of live births per 1000 women in yStat. To do this, press [2nd] [STAT] [F2] to get into the statistical list-editor screen. Clear any data in xStat and yStat by positioning the cursor over the name of the list to be cleared. Then press [CLEAR] [ENTER]. Once both lists are cleared enter the data as we have done here. The lists are too long to be seen on one screen. Use the up- and down-arrow keys to scroll through the lists.

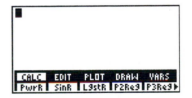   

Then press [EXIT] to return to the home screen. Press [2nd] [STAT] [F1] [MORE] and look at the menu on the bottom of the screen. To select the QUADRATIC REGRESSION feature, "P2Reg," from the STATISTICAL CALCULATIONS menu, press [F4].

Now to tell the calculator which lists contain the data, press [2nd] [LIST] [F3] and look at the menu at the bottom of the screen. The data is in xStat and yStat. The list of named lists can vary from calculator to calculator, depending on who used it and what they named the lists. You need to find "xStat" and "yStat" in your calculator. You may need to press [MORE] to find the lists. Remember to separate their names using a comma.

Before the regression equation is found, it is possible to select a y-variable to which it will be stored on the equation-editor screen. To do this, press [,] [2nd] [alpha] Y. (Y is the alphabetic character associated with the [0] key.) Now press [1]. Finally, press [ENTER] to display the coefficients of a quadratic function of the form

$f(x) = ax^2 + bx + c$. Rounding the coefficients to two decimal places, we obtain the

function $f(x) = -0.49x^2 + 25.95x - 238.49$. The right-arrow key can be used to reveal all three coefficients.

We also want to see the data plotted. Press [2nd] [STAT] and look at the menu at the bottom of the screen. Press [F3] to plot. Choose "Plot 1" by pressing [F1].

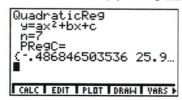

Then make your screen look like this one.

To graph, press [GRAPH] and look at the menu. Press [F3] to choose "ZOOM."

Press [MORE] and look at the menu. To have the calculator set the window, press [F5] to choose "ZDATA." To clear the menu from the screen, press [CLEAR].

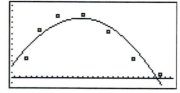

(b)   Fit a cubic function to the data using the REGRESSION feature on the calculator. Then make a scatter plot of the data and graph the cubic function with the scatter plot.

Once the data are entered, fit a cubic function to it by pressing [2nd] [STAT] [F1] [MORE] [F5]. These keystrokes select the CUBIC REGRESSION feature, "P3Reg," from the STATISTICAL CALCULATIONS menu.

Now press [2nd] [LIST] [F3]. These keystrokes take you to the list of all the named lists in your calculator. As before, find "xStat" and "yStat." Press the appropriate F keys to select "xStat" and "yStat." Remember to separate them using a comma.

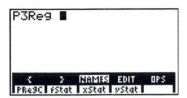   

Tell the calculator to put the equation to the right of "y2 =" on the equation-editor screen by pressing [,] [2nd] [alpha] Y [2].

Press [ENTER] [EXIT] [EXIT] and the calculator displays the coefficients of a cubic function of the form $f(x) = ax^3 + bx^2 + cx + d$. Use the right-arrow key to see all three coefficients.

Rounding the coefficients to two decimal places we have $f(x) = 0.03x^3 - 3.22x^2 + 101.18x - 886.93$.

Before graphing the cubic function we need to turn off the quadratic function. Press [GRAPH] [F1] to see both the quadratic and cubic equations on the equation-editor screen. Position the cursor to the right of "y1 =" and press [F5]. This deselects the function, or turns it off. When we graph, it will not be shown. Since Plot 1 is already turned on (notice "Plot 1" is highlighted), and we already have a suitable window, press [GRAPH] [F5] [CLEAR] to see the data points and the cubic function graphed together.

      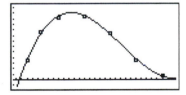

(c)   Which function seems to fit the data better?

The graph of the cubic function appears to fit the data better than the graph of the quadratic function. Remember that a prediction is only an approximation.

(d)   Use the function from part (c) to estimate the average number of live births by women of ages 20 and 30.

We can use any of the methods described earlier in this chapter to evaluate the function. We will use the FORECAST feature from the STATISTICS menu. Press [2nd] [STAT] [MORE] [F1]. The forecast screen will be displayed with the cursor beside "x =," press [2] [0] [ENTER]. The cursor moves beside "y =." Press [F5] to select "SOLVE." We estimate that there will be about 100 live births per 1000 20-year-old women. Press [▲] to position the cursor beside "x =" again and press [3] [0] [ENTER] to change the 20 to 30. With the cursor beside "y =," press [F5] to select "SOLVE." We estimate that there will be about 97 live births per 1000 30-year-old women.

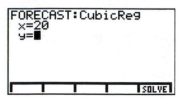
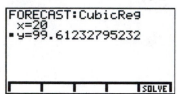
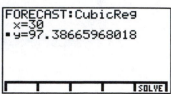

Chapter 1
Differentiation

ZOOM IN FEATURE

The ZOOM IN feature from the GRAPH ZOOM menu can be used to enlarge a portion of a graph.

Section 1.2, Technology Connection, page 123   Verify graphically that $\lim\limits_{x \to 0} \dfrac{\sqrt{x+1}-1}{x} = 0.5$.

First, graph $y1 = \dfrac{\sqrt{x+1}-1}{x}$ in a window that shows the portion of the graph near $x = 0$. One good choice for the window is [-2, 5, -1, 2]. This function requires several sets of parenthesis. Look at the equation-editor screen below to make sure you have used the parenthesis correctly. Press GRAPH F2 to change the window settings. Be sure the plots are turned off and that all other entries on the equation-editor screen are cleared. With the window set and the function entered, press GRAPH F5 to see the graph displayed. Press F4 and use the ◀ and ▶ keys to move the cursor to a point on the curve near $x = 0$.

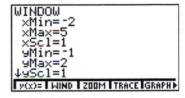

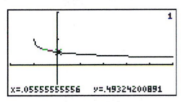

Now select the ZOOM IN feature, "ZIN," from the GRAPH ZOOM menu by pressing GRAPH F3 F2 ENTER. This enlarges the portion of the graph near $x = 0$. We can now press GRAPH F4 and use the left- and right-arrow keys to trace the curve near $x = 0$.

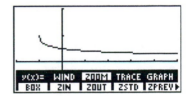
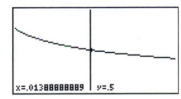
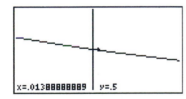

The ZOOM IN feature can be used as many times as desired in order to verify that the limit as x approaches 0 is .5.

THE TABLE FEATURE

For an equation entered in the equation-editor screen, a TABLE of x- and y-values can be displayed.

Section 1.2, Technology Connection, page 123   Verify graphically that $\lim\limits_{x \to 0} \dfrac{\sqrt{x+1}-1}{x} = 0.5$.

First, graph $y1 = \dfrac{\sqrt{x+1}-1}{x}$ in a window that shows the portion of the graph near $x = 0$. One good choice

for the window is [-2, 5, -1, 2]. Press $\boxed{\text{TABLE}}$. Press $\boxed{\text{F2}}$ and use the settings below in the middle. Press $\boxed{\text{F1}}$ and the

TABLE will be displayed.

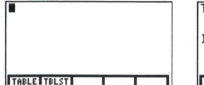  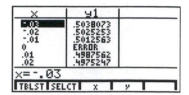

This is a way to verify that the limit is .5.

THE nDer FEATURE

The NUMERICAL DERIVATIVE feature, "nDer(," from the CALCULUS menu can be used to evaluate a
numerical derivative on the home screen. Remember that the numerical derivative is the slope of the tangent line to
a curve at a specific point.

Section 1.4, Technology Connection, page 144   For the function $f(x) = x(100-x)$, find $f'(70)$.

We will use the NUMERICAL DERIVATIVE feature from the CALCULUS menu. Go to the home screen
and select this operation by pressing $\boxed{\text{2nd}}$ $\boxed{\text{CALC}}$ $\boxed{\text{F2}}$ to select "nDer(" from the CALCULUS menu. ($\boxed{\text{CALC}}$ is the
second function associated with the $\boxed{\div}$ key.) Now enter the function, the variable, and the value at which the
derivative is to be evaluated, all separated by commas. Press $\boxed{\text{x-VAR}}$ $\boxed{(}$ $\boxed{1}$ $\boxed{0}$ $\boxed{0}$ $\boxed{-}$ $\boxed{\text{x-VAR}}$ $\boxed{)}$ $\boxed{,}$ $\boxed{\text{x-VAR}}$ $\boxed{,}$ $\boxed{7}$ $\boxed{0}$ $\boxed{)}$.
Press $\boxed{\text{ENTER}}$ and the numerical derivative is given. Note that the graphing calculator supplies a left parenthesis after
"nDer" and we supply the right parenthesis after entering 70.

 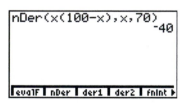

If the function is entered on the equation-editor screen as in the screen below on the left, we can evaluate
the numerical derivative by going to the home screen and pressing $\boxed{\text{2nd}}$ $\boxed{\text{CALC}}$ $\boxed{\text{F2}}$ $\boxed{\text{2nd}}$ [alpha] Y $\boxed{1}$ $\boxed{,}$ $\boxed{\text{x-VAR}}$ $\boxed{,}$ $\boxed{7}$ $\boxed{0}$
$\boxed{)}$. Press $\boxed{\text{ENTER}}$ to see the derivative.

 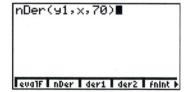 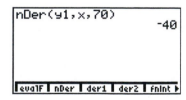

We see that $f'(70) = -40$. Since this is the value of the numerical derivative at $x = 70$, we know that the

slope of the tangent line to the curve at $x = 70$ is -40.

THE TanLn FEATURE; On the Home Screen

We can draw a line tangent to a curve at a given point using the TANGENT LINE feature from the GRAPH DRAW menu.

Section 1.4, Technology Connection, page 144 Draw the line tangent to the graph of $f(x) = x(100 - x)$, at $x = 70$.

First, graph $y1 = x(100 - x)$ in a window that shows the portion of the curve near $x = 70$. One good window choice is shown below. Be sure to clear any functions that were previously entered and turn off the plots.

```
WINDOW
 xMin=-10
 xMax=100
 xScl=10
 yMin=-10
 yMax=3000
↓yScl=1000■
[ y(x)= | WIND | ZOOM | TRACE | GRAPH ▶
```

Now press [2nd] [QUIT] to go to the home screen. ([QUIT] is the second function associated with the [EXIT] key.) Here we will select the TANGENT LINE feature from the GRAPH DRAW menu and instruct the calculator to draw the line tangent to the graph of $y1$ at $x = 70$. Press [GRAPH] [MORE] [F2] [MORE] [MORE] [MORE] [F2] [2nd] [alpha] Y [1] [,] [7] [0] [)] [ENTER]. Press [CLEAR] to clear the menu from the bottom of the screen. Note that the calculator supplies a left parenthesis after "TanLn" and we close the parentheses with a right parenthesis after entering 70.

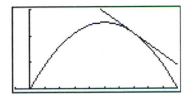

To clear the drawing from the graph screen, use the CLEAR DRAWING command, "CLDRW," from the GRAPH DRAW menu. Press [GRAPH] [MORE] [F2] [MORE] [MORE] [F1].

See page 112 of this manual for a procedure that draws a tangent line directly from the graph screen.

THE $\dfrac{dy}{dx}$ FEATURE

The NUMERICAL DERIVATIVE feature, "dy/dx," from the GRAPH MATH menu can be used to evaluate a numerical derivative on the graph screen. Remember that the numerical derivative is the slope of the tangent line to a curve at a specific point. We can find the derivative of a function at a specific point directly from the graph screen.

Section 1.5, Technology Connection, page 149 For the function $f(x) = x\sqrt{4 - x^2}$, find $\dfrac{dy}{dx}$ at a specific point.

First graph $y1 = x\sqrt{4 - x^2}$. We will use the window [-3, 3, -4, 4]. Then select "dy/dx" from the GRAPH MATH menu by pressing [GRAPH] [MORE] [F1] [F2]. We see the graph with a cursor positioned on the x-value that is the x-coordinate of the midpoint of the x-axis.

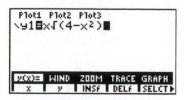 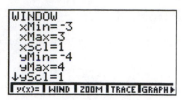 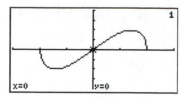

To find the value of dy/dx at a specific point either move the cursor to the desired point or key in the point's x-coordinate. For example, press [GRAPH] [MORE] [F1] [F2], as explained above, and move the cursor to the point (1.2380952381, 1.9446847395). Now press [ENTER] and the derivative is displayed. We see that $dy/dx = 0.5947897471$ at this point.

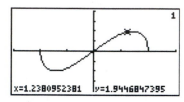 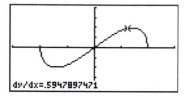

To evaluate dy/dx for $x = 1$, select "dy/dx" from the GRAPH MATH menu by pressing [GRAPH] [MORE] [F1] [F2] and then press [1] [ENTER].

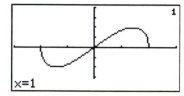 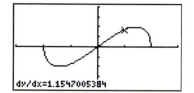

We see that $dy/dx = 1.1547$, when $x = 1$.

THE TANGENT FEATURE; On the Graph Screen

Section 1.5, Technology Connection, page 149 Draw the line tangent to the graph of $f(x) = x\sqrt{4 - x^2}$ at a specific point.

First, graph $f(x) = x\sqrt{4 - x^2}$ in an appropriate viewing window. Here we are using [-3, 3, -4, 4]. Now select "TANLN" from the GRAPH MATH menu by pressing [GRAPH] [MORE] [F1] [MORE] [MORE] [F1] use the left- and right-arrow keys to move the cursor to the desired point. Press [ENTER] and see the tangent line drawn at that point. The slope of this tangent line is also displayed.

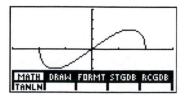 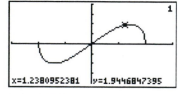 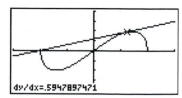

Use "CLDRW" (clear drawing) from the GRAPH DRAW menu to clear the graph of the tangent line from the graph screen. If the graph and the tangent line are still displayed, press [GRAPH] [MORE] [F2] [MORE] [MORE] [F1]. The graph is redrawn without the tangent line. Press [CLEAR] to remove the menus.

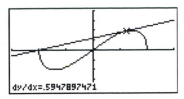 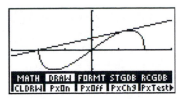 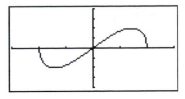

We can enter the *x*-coordinate of the point where we want the tangent line to be drawn. For example,
To graph the line tangent to the curve at $x = 1$, from the graph screen press [GRAPH] [MORE] [F1] [MORE] [MORE] [F1] as
before to select "TANLN." Then press [1] [ENTER] to enter 1 as the *x*-coordinate.

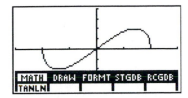 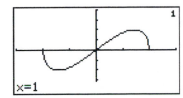 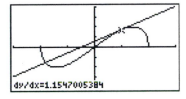

The tangent line to the curve and the slope of the tangent line at that point are both displayed.

Use "CLDRW" (clear drawing) from the GRAPH DRAW menu to clear the graph of the tangent line from
the graph screen. If the graph and the tangent line are still displayed, press [GRAPH] [MORE] [F2] [MORE] [MORE] [F1]
[CLEAR] and the graph will be redrawn without the tangent line.

ENTERING THE SUM OF TWO FUNCTIONS

Section 1.5, Technology Connection, page 152 Given that $y1 = x(100 - x)$, $y2 = x\sqrt{100 - x^2}$, and $y3 = y1 + y2$,
find the derivative of each of the three functions at $x = 8$. First press [GRAPH] [F1] to go to the equation-editor screen.
Clear any entries that are present and turn off the plots. Then enter $y1 = x(100 - x)$ and $y2 = x\sqrt{100 - x^2}$. To enter
$y3 = y1 + y2$, position the cursor to the right of "y3 =" and press [F2] [1] [+] [F2] [2].

Press [2nd] [QUIT] to go to the home screen. We will use the NUMERICAL DERIVATIVE feature, "nDer("
to find the derivative of each of the three functions at $x = 8$ on the home screen. Select this operation by pressing
[2nd] [CALC] [F2]. These keystrokes copy "nDer(" to the home screen. Now enter y1 by pressing [2nd] [alpha] Y [1] [,],
the variable by pressing [x-VAR] [,], and the value at which the derivative is to be evaluated by pressing [8] [)]. Notice
that the fields are separated by commas. Press [ENTER] and the numerical derivative is given. Note that the graphing
calculator supplies a left parenthesis after "nDer" and we supply the right parenthesis after entering 8.

  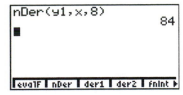

While still on the home screen notice the menu items which are still displayed on the bottom of the screen. To do another numerical derivative simply press [F2]. Now enter y2 by pressing [2nd] [alpha] Y [2] [,], the variable by pressing [x-VAR] [,], and the value at which the derivative is to be evaluated by pressing [8] [ )]. Press [ENTER] and the numerical derivative is given.

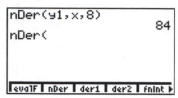   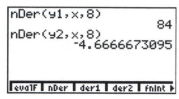

For the third numerical derivative, we press [2nd] [ENTRY] to bring back the last entry on the home screen. ([ENTRY] is the second function associated with the [ENTER] key.)   Use the left-arrow key to position the cursor over the 2 in y2 and change it into a 3. Then press [ENTER] and the numerical derivative is given.

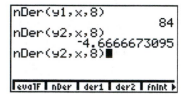 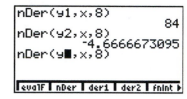 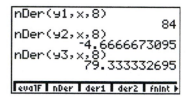

Notice that the third derivative is the sum of the first two.

DESELECTING FUNCTIONS; GRAPH STYLES

Several functions can be graphed at the same time and they can be graphed using different graph styles. Functions can be deselected on the equation-editor screen so that they remain there, but will not be graphed.

Section 1.6, Technology Connection and Example 5, page 162   We have the original function, $y1 = \dfrac{x^2 - 3x}{x - 1}$, the

derivative found in Example 5 on page 162 in the text book, $y2 = \dfrac{x^2 - 2x + 3}{(x-1)^2}$, and the derivative the calculator will

graph, $y3 = nDer(y1, x, x)$. If the graphs of y2 and y3 coincide, the answer in Example 5 is verified by the calculator.

First enter $y1 = \dfrac{x^2 - 3x}{x - 1}$ and $y2 = \dfrac{x^2 - 2x + 3}{(x-1)^2}$. Now enter $y3 = nDer(y2, x, x)$. Position the cursor immediately to the

right of "y3=" and press [2nd] [CALC] [F2] [GRAPH] [F1] [F2] [1] [,] [F1] [,] [F1] [ )].

Since we want to see only the graphs of y2 and y3, we will deselect y1. Position the cursor to the right of the "=" in "y1=" and press [F5]. This deselects the function, or turns it off so it will not be graphed.

To select y1 again, position the cursor anywhere in the function y1 and press [F5]. The equals sign will be highlighted now, indicating that he function is selected. For now, leave it deselected.

With y1 deselected, we can graph y2 and y3 using different graph styles to determine whether the graph coincide. When the graphing calculator is in CONNECTED mode, equations are graphed with a solid line. You can see the graph style icon to the left of each of the *y*-variable names looks like a solid line.

Make sure the calculator is in SEQUENTIAL mode before proceeding. Press [GRAPH] [MORE] [F3] and make sure "SeqG" is highlighted. We will keep the solid line graph style for the graph of y2 and select the path style for the graph of y3. After the graph of y2 is drawn, a circular cursor will trace over the graph that is already displayed. To select the path style for y3, position the cursor somewhere in the y3 function and press [GRAPH] [F1] [MORE] and select "STYLE" by pressing [F3] four times, until you see "⊕."

Press [GRAPH] [F3] [F4]. The cursor that resembles a ball moves along the first graph. We could also examine the TABLE to verify the results. Below are the table settings and the table these settings produce.

Both verify that the derivative found in Example 5 is correct.

Chapter 2
Applications of Differentiation

THE FMAX and FMIN FEATURES

Section 2.1, Technology Connection, page 210   Use the FUNCTION MAXIMUM, "FMAX," and FUNCTION

MINIMUM, "FMIN," features from the GRAPH MATH menu to approximate the relative extrema of

$f(x) = -0.4x^3 + 6.2x^2 - 11.3x - 54.8$ on the graph screen.

First, graph $y1 = -0.4x^3 + 6.2x^2 - 11.3x - 54.8$ in a window that displays the relative extrema of the

function. The window settings shown below were found using trial and error. Observe that a relative maximum

occurs near x = 10 and a relative minimum occurs near x = 1.

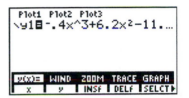 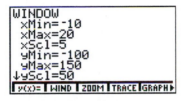 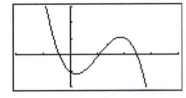

To find the relative maximum, first press GRAPH MORE F1 F5 to select "FMAX" from the GRAPH

MATH menu. We are prompted to select a left bound for the relative maximum. This means that we must choose

an *x*-value that is to the left of the *x*-coordinate of the point where the relative maximum occurs. This can be done

by using the left- and right-arrow keys to move the cursor to a point to the left of the relative maximum or by keying

in an appropriate value.

 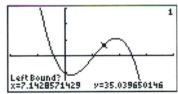

Once this is done, press ENTER. Notice that a triangular marker has appeared to mark the left bound. Now

we are prompted to select a right bound. Either move the cursor to a point to the right of the relative maximum or

key in an appropriate value.

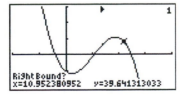

Press ENTER. Notice another triangular marker has appeared to mark the right bound. These two markers

provide boundaries for the calculator. Finally we are prompted to guess the *x*-value at which the relative maximum

occurs. Move the cursor near the relative maximum point or key in an *x*-value. Press ENTER a third time. We see

that the relative maximum function value is approximately 54.61 when $x \approx 9.32$.

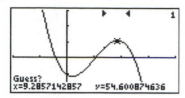
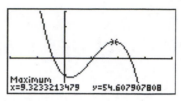

To find the relative minimum, select "FMIN" from the GRAPH MATH menu by pressing GRAPH MORE F1 F4 ENTER. Select left and right bounds for the relative minimum and guess the x-value at which it occurs as described on the previous page.

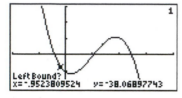
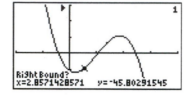
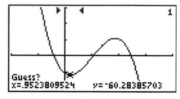

We see that a relative minimum function value of approximately -60.30 occurs when $x \approx 1.01$.

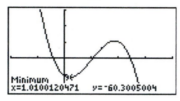

To summarize, a relative maximum function value is approximately 54.61 when $x \approx 9.32$ and a relative minimum function value of approximately -60.30 occurs when $x \approx 1.01$.

THE fMax and fMin FEATURES

The FUNCTION MAXIMUM, "fMax(," and FUNCTION MINIMUM, "fMin(," features from the CALCULUS menu can be used on the home screen to calculate the x-values at which relative maximum and minimum values of a function occur over a specified closed interval.

Section 2.1, Technology Connection, page 210 Use "fMax(" and "fMin(" from the CALCULUS menu to approximate the relative extrema of $f(x) = -0.4x^3 + 6.2x^2 - 11.3x - 54.8$ on the home screen.

First enter the function as y1 and graph it as shown below. Observe that a relative maximum occurs in the closed interval [5, 15]. There are other intervals we could use. Keep in mind that the larger the interval, the longer it takes the calculator to return an x-value.

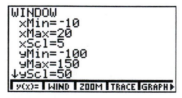
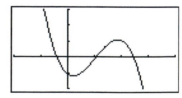

Now press 2nd [QUIT] to go to the home screen and then press 2nd [CALC] MORE and look at the menu at the bottom of the screen. To select "fMax(" from the CALCULUS menu press F2. Enter the name of the function, the variable, and the left and right endpoints of the interval on which the relative maximum occurs by pressing 2nd [alpha] Y 1 , x-VAR , 5 , 1 5). Press ENTER. The relative maximum occurs when $x \approx 9.32332128289$. To find the relative maximum value of the function, we evaluate the function for this value of x. Press 2nd [alpha] Y

1 ((2nd [ANS]) ENTER . ([ANS] is the second function associated with the (-) key.) The keystrokes 2nd [ANS] cause

the calculator to use the previous answer, 9.32332128289, as the value for x in y1.

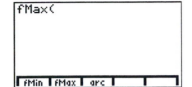 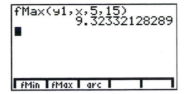 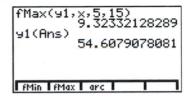

We also observe that a relative minimum occurs in the interval [-5, 5]. Again, there are other intervals we

could choose. To find the relative minimum in this interval, first press 2nd [CALC] MORE F1 to select "fMin(" from

the CALCULUS menu. Then enter the name of the function, the variable, and the endpoints of the interval. Press

2nd [alpha] Y 1 , x-VAR , (-) 5 , 5) . Press ENTER . The relative minimum occurs when $x \approx 1.01001202891$.

To find the relative minimum value of the function, we evaluate the function for this value of x. Press 2nd [alpha] Y

1 ((2nd [ANS]) ENTER . ([ANS] is the second function associated with the (-) key.) The keystrokes 2nd [ANS] cause

the calculator to use the previous answer, 1.01001202891, as the value for x in y1.

 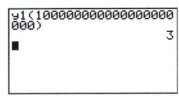

To summarize, a relative maximum function value is approximately 54.61 when $x \approx 9.32$ and a relative

minimum function value of approximately -60.30 occurs when $x \approx 1.01$.

LIMITS AT INFINITY and HORIZONTAL ASYMPTOTES

Section 2.3, Technology Connection, page 234 Use function notation to verify that $\lim\limits_{x \to \infty} \dfrac{3x-4}{x} = 3$ and find the

function's horizontal asymptote.

With $y1 = \dfrac{3x-4}{x}$, go to the home screen and press 2nd [alpha] Y 1 (. We use a really large number, since

we are trying to approach infinity. Press 1 0 continue pressing 0 until you have 15 or more of them. Then press

) ENTER .

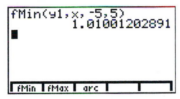 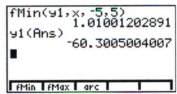

We see that the limit at infinity is 3. Also, the horizontal asymptote is the line $y = 3$.

THE FMAX and FMIN FEATURES

Section 2.4, Technology Connection and Example 1, page 248   Use the FUNCTION MAXIMUM, "FMAX,"
and FUNCTION MINIMUM, "FMIN," features in the GRAPH MATH menu to approximate the absolute extrema
of $f(x) = x^3 - 3x + 2$ over the interval $\left[-2, \dfrac{3}{2}\right]$ on the graph screen.

First graph $y1 = x^3 - 3x + 2$ in a window with $xMin = -2$ and $xMax = \dfrac{3}{2}$. Over this interval an absolute

maximum occurs near $x = -1$ and an absolute minimum occurs near $x = 1$ and $x = -2$. Since we have a window

which displays the entire interval we are interested in, it is obvious that the endpoint $x = \dfrac{3}{2}$ is not the x-coordinate of

an absolute maximum or minimum point over this interval.

 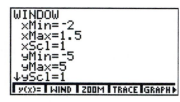 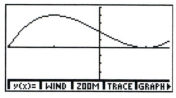

To find the absolute maximum, press [GRAPH] [MORE] [F1] [F5] to select "FMAX" from the GRAPH MATH
menu. We are prompted to select a left bound for the absolute maximum. This means that we must choose an x-
value that is to the left of the x-coordinate of the point where the absolute maximum occurs. This can be done by
using the left- and right-arrow keys to move the cursor to a point to the left of the absolute maximum or by keying in
an appropriate value.

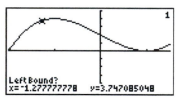

Once this is done, press [ENTER]. Notice that a triangular marker has appeared to mark the left bound. Now
we are prompted to select a right bound. Move the cursor to a point to the right of the absolute maximum or key in
an appropriate value.

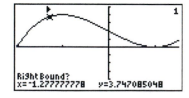 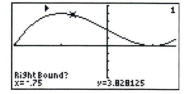

Press [ENTER]. Notice another triangular marker has appeared to mark the right bound. These two markers
provide boundaries for the calculator. Finally we are prompted to guess the x-value at which the absolute maximum
occurs. Move the cursor near the absolute maximum point or key in an x-value. Press [ENTER] a third time. We see
that the absolute maximum function value is 4 when $x = -1$.

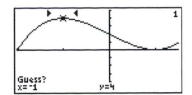 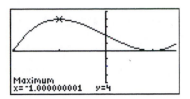

To find the absolute minimum, select "FMIN" from the GRAPH MATH menu by pressing [GRAPH] [MORE] [F1] [F4]. Select left and right bounds for the absolute minimum and guess the x-value at which it occurs as described on the previous page.

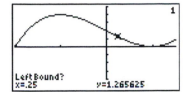 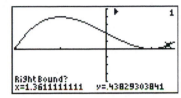 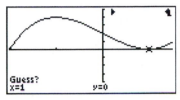

On this interval, we have an absolute minimum function value of 0 occurs when $x = 1$.

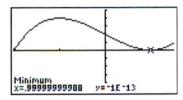

NOTE: The number -1E-13 is in scientific notation on the calculator. It means -1×10^{-13}. This number written as a decimal is -0.0000000000001. This is very nearly 0.

We are also obligated to check the endpoints of the interval $\left[-2, \dfrac{3}{2} \right]$. To do this we use the EVALUATE feature from the GRAPH menu. With the graph of the function displayed, press [GRAPH] [MORE] [MORE] [F1] to choose "EVAL" from the GRAPH menu. We must enter the x-value, -2, and then press [ENTER]. A second absolute minimum function value of 0 occurs when $x = -2$.

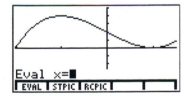 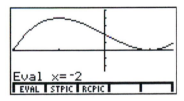 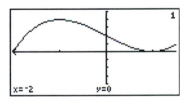

We can see that the other endpoint is neither a maximum nor a minimum, but we repeat the procedure described above for completeness. Note that if we have just used the EVALUATE feature, we only have to enter the next x-value we want to evaluate; we do not have to return to the menu.

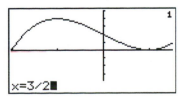 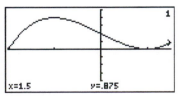

To summarize, the absolute maximum value of the function over the interval is 4 when $x = -1$. The absolute minimum value of the function over the interval is 0 when $x = 1$ or when $x = -2$.

THE fMax and fMin FEATURES

The FUNCTION MAXIMUM, "fMax(," and FUNCTION MINIMUM, "fMin(," features from the CALCULUS menu can be used on the home screen to calculate the x-values at which relative maximum and minimum values of a function occur over a specified closed interval.

Section 2.4, Technology Connection and Example 1, page 248   Use "fMax(" and "fMin(" from the CALCULUS menu to approximate the absolute extrema of $f(x) = x^3 - 3x + 2$ over the interval $\left[-2, \dfrac{3}{2}\right]$ on the home screen.

First enter the function as y1 and graph it as shown below.

 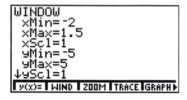 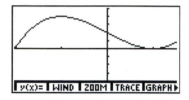

Press [2nd] [QUIT] to go to the home screen and press [2nd] [CALC] [MORE] [F2] to select "fMax(" from the CALCULUS menu. Enter the name of the function, the variable, and the left and right endpoints of the interval on which the absolute maximum occurs by pressing [2nd] [alpha] Y [1] [,] [x-VAR] [,] [(-)] [2] [,] [3] [÷] [2] [)]. Press [ENTER] to find that the absolute maximum occurs when $x = -1$. To find the absolute maximum value of the function, we evaluate the function for this value of x. Press [2nd] [alpha] Y [1] [(] [2nd] [ANS] [)] [ENTER]. ([ANS] is the second function associated with the [(-)] key.) The keystrokes [2nd] [ANS] cause the calculator to use the previous answer, -1, as the value for x in y1. The absolute maximum value of the function is 4 when $x = -1$.

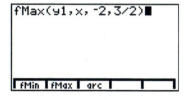 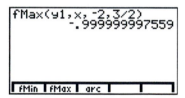 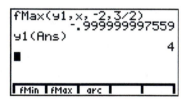

We also observe that two absolute minimums occur in the interval $\left[-2, \dfrac{3}{2}\right]$. We will have to use different intervals in order to find both of them. The calculator will find one of them if we use the entire interval, so press [2nd] [CALC] [MORE] [F1] to select "fMin(" from the CALCULUS menu. Then enter the name of the function, the variable, and the endpoints of the interval. Press [2nd] [alpha] Y [1] [,] [x-VAR] [,] [(-)] [2] [,] [3] [÷] [2] [)]. Press [ENTER]. One of the absolute minimums occurs when $x = 1$. To find the absolute minimum value of the function, we evaluate the function for this value of x. Press [2nd] [alpha] Y [1] [(] [2nd] [ANS] [)] [ENTER]. ([ANS] is the second function associated with the [(-)] key.) The keystrokes [2nd] [ANS] cause the calculator to use the previous answer, 1, as the value for x in y1.

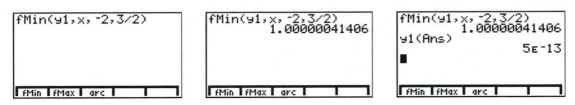

NOTE: The number 5E-13 is in scientific notation on the calculator. It means 5×10^{-13}. This number written as a decimal is 0.0000000000005. This is very nearly 0.

To find the second absolute minimum, the interval must be changed. The new interval must not contain the x-value we just found, that is to say, 1 can not be in the new interval. Press [2nd] [CALC] [MORE] [F1] to select "fMin(" from the CALCULUS menu. Then enter the name of the function, the variable, and the endpoints of the interval. We are using the closed interval from -2 to -1. Press [2nd] [alpha] Y [1] [,] [x-VAR] [,] [(-)] [2] [,] [(-)] [1] [)]. Press [ENTER]. The second absolute minimum occurs when $x = -2$. To find the absolute minimum value of the function, we evaluate the function at this x-value. Press [2nd] [alpha] Y [1] [(] [2nd] [ANS] [)] [ENTER]. ([ANS] is the second function associated with the [(-)] key.) The keystrokes [2nd] [ANS] cause the calculator to use the previous answer, -2, as the value for x in y1.

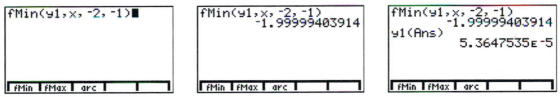

NOTE: The number 5.3647535E-5 is scientific notation on the calculator. It means 5.3647535×10^{-5}. This number written as a decimal is 0.000053647535. This is very nearly 0.

To summarize, the absolute maximum value of the function over the interval is 4 when $x = -1$. The absolute minimum value of the function over the interval is 0 when $x = 1$ or $x = -2$.

THE LISTS (as a spread sheet)

The lists can be used like a spread sheet to solve problems which involve mathematical formulas.

Section 2.5, Technology Connection, page 260

1.   Use the lists to complete the table:

x	$y = 20 - x$	$A = x(20 - x)$
0		
4		
6.5		
8		
10		
12		
13.2		
20		

Using the lists in the calculator, this table becomes:

x LENGTH	$y = 20 - x$ WIDTH = 20 – LENGTH	$A = x(20 - x)$ AREA = LENGTH * WIDTH
0		
4		
6.5		
8		
10		
12		
13.2		
20		

Press [2nd] [STAT] [F2] and you will be on the statistical list-editor screen. Move the cursor to the top row and then to the right until you see a screen like the one below. We will name the lists we use. It does not matter what lists are already named in your calculator.

We are going to name the list LENGTH. L, E, N, G, T, and H are the alphabetic characters associated with the [7] [^] [9] [EE] [–] [(] keys, respectively. We can do this since the calculator is in ALPHABETIC mode as indicated by the ⓐ in the upper right corner or the screen. After the name is typed, press [ENTER] and the list is named.

Press [▼] to get into the list. Now enter 0, 4, 6.5, 8, 10, 12, 13.2, and 20. After entering each value, press [ENTER] or [▼] to drop down to the next line and make the next entry in the list. This list is too long to be seen on a single screen. Use the [▲] and [▼] keys to scroll up and down in the list and view all the data.

With the data in, move the cursor up to highlight "LENGTH" and press [▶] to name the next list. Notice the ⓐ in the upper right corner. It indicates that the calculator is in ALPHABETIC mode. As you type the name of the list, the icon goes away, but the calculator is remains in ALPHABETIC mode. We name the list WIDTH by pressing the keys [3] [)] [TAN] [–] [(]. Press [ENTER].

Press ENTER and notice that the cursor has moved to the right of "WIDTH =" immediately above the menu at the bottom of the screen. We can use a formula and define the value of each entry in the list. The formula must be enclosed in quotation marks. Press F4 2 0 − F3. If "LENGTH" is not displayed, press MORE until you find it. Then press the appropriate F key to choose it. Press 2nd [M4] to supply the quotation mark. ([M4] is the second function associated with the F4 key.) Press ENTER and the list will fill in.

Press ▲ to get to the list's name and press ▶ to get to another unnamed list. As before enter the name of the new list. Press LOG 5 ^ LOG to name the list AREA. Press ENTER to name the list. Press ENTER again to move the cursor to be bottom above the menu.

To enter the formula, press F3. If "LENGTH" is not displayed, press MORE until you find it. Then press the appropriate F key to choose it. Press ×. If "WIDTH" is not displayed, press MORE until you find it. Then press the appropriate F key to choose it. Press ENTER. The list will fill in. Use the down-arrow to scroll and see the rest of the list.

We can use these to fill in the original table. Here is the table filled in completely.

x LENGTH	$y = 20 - x$ WIDTH = 20 − LENGTH	$A = x(20-x)$ AREA = LENGTH * WIDTH
0	20	0
4	16	64
6.5	13.5	87.75
8	12	96
10	10	100
12	8	96
13.2	6.8	89.76
20	0	0

From this table, we see that the maximum area is 100 square feet. The maximum area occurs when the length and width are both 10 feet.

2.   Graph $A(x) = x(20 - x)$ over the interval [0, 20].

Press GRAPH F1 and clear existing entries and turn off the plots. The equation becomes $y1 = x(20 - x)$. The y-value in the calculator is the area and the x-value is the x-value in the equation. Below are the equation-editor and window screens. Press GRAPH F5 CLEAR to see the area function graphed on the interval we chose.

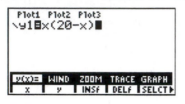 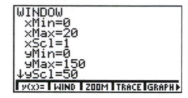 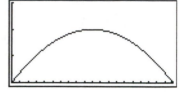

3. Estimate a maximum value of the function and tell the x-value which generates it.

To find the maximum, first press GRAPH MORE F1 F5 to select "FMAX" from the GRAPH MATH menu. We are prompted to select a left bound for the maximum. This means that we must choose an x-value that is to the left of the x-coordinate of the point where the absolute maximum occurs. This can be done by using the ◄ and ► keys to move the cursor to a point to the left of the absolute maximum or by keying in an appropriate value.

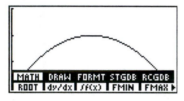 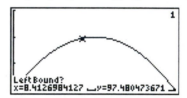

Press ENTER. Notice that a triangular marker has appeared to mark the left bound. Now we are prompted to select a right bound. Move the cursor to a point to the right of the absolute maximum or key in an appropriate value.

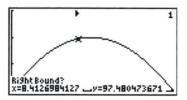 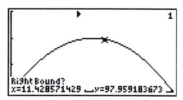

Press ENTER. Notice another triangular marker has appeared to mark the right bound. These two markers provide boundaries for the calculator. Finally we are prompted to guess the x-value at which the absolute maximum occurs. Move the cursor near the absolute maximum point or key in an x-value. Press ENTER a third time. We see that the maximum area is 100 square feet when $x = 10$. The rectangle is 10 feet by 10 feet. That makes it a special rectangle. It is actually a square.

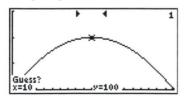 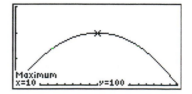

OPTIMIZATION APPLICATION

Section 2.5, Example 2, page 261 *Maximizing Volume* From a thin piece of cardboard 8 inches by 8 inches, square corners are cut out so that the sides can be folded up to make a box. What dimensions will yield a box of maximum volume? What is the maximum volume? (See diagrams in your text.)

The volume of a box = length • width • height. Here we have $V = (8-2x)(8-2x)x$. We will graph this function and find the maximum point on a chosen interval. On the equation-editor screen clear all functions and turn off the plots. Enter the volume as $y1 = (8-2x)(8-2x)x$. Since the original material was 8" by 8" and equal squares are being cut out at the corners and the sides folded up, the domain must be in the closed interval between 0 and 4. Use the window below and then graph the function. Press CLEAR to remove the menu from the graph.

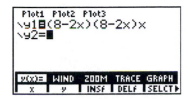 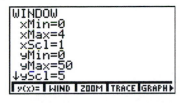 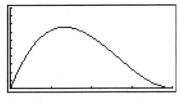

It is worth noting that since this is a cubic function it actually has no absolute maximum; however we are using it to model a real world problem with constraints and therefore are working on a closed interval. There is an absolute maximum over every closed interval. Here we use "FMAX" from the GRAPH MATH menu to find the maximum volume. Press GRAPH MORE F1 F5 and answer the questions on the screens by moving the cursor and pressing ENTER as indicated below.

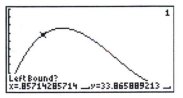 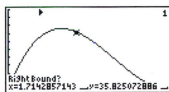 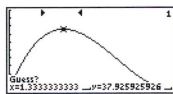

Press ENTER and the maximum point is revealed.

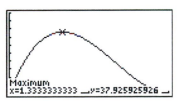

Your answer may vary depending upon how you bound and how you guess. Here $x = 1.333333...$ or $x = \dfrac{4}{3}$.

To find the other dimensions, go to the home screen by pressing 2nd [QUIT]. Press 8 − 2 (4 ÷ 3) and then press ENTER to see the other dimensions of the box.

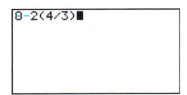 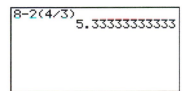

The length and width of the box are equal. They are both $5\frac{1}{3}$ inches. The height is $1\frac{1}{3}$ inches or $\frac{4}{3}$ inches.

To find the volume, press [2nd] [ANS] [×] [2nd] [ANS] [×] [(] [4] [÷] [3] [)]. This takes the previous answer on the home

screen, $5\frac{1}{3}$, and multiplies it by itself and then by the height, $\frac{4}{3}$. The decimal approximation of the volume is given.

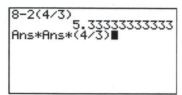   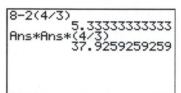

To write this volume as a mixed number, subtract the whole number, 37, and then press [2nd] [MATH] [F5]

[MORE] [F1] [ENTER] to turn the decimal part into a fraction. The volume is $37\frac{25}{27}$ cubic inches.

```
              37.9259259259
Ans-37
              .925925925926
Ans▶Frac
                     25/27

 NUM  PROB  ANGLE  HYP   MISC
▶Frac   2   ▶Eval   ×√   ºval
```

THE LISTS (as a spread sheet)

The lists can be used like a spread sheet to solve problems which involve mathematical formulas.

Section 2.5, Technology Connection, page 268 Use the calculator to fill in the table:

Lot size, x	Number of Recorders, $\dfrac{2500}{x}$	Average Inventory, $\dfrac{x}{2}$	Carrying costs, $10 \cdot \dfrac{x}{2}$	Cost of each order, $20+9x$	Reorder Costs, $(20+9x)\dfrac{2500}{x}$	Total Inventory Costs, $C(x)=10 \cdot \dfrac{x}{2}+(20+9x)\dfrac{2500}{x}$
2500	1	1250	$12,500	$22,200	$22,520	$35,020
1250	2	625	$6,250	$11,270	$22,540	
500	5	250	$2,500	$4,520		
250	10	125				
167	15	84				
125	20					
100	25					
90	28					
50	50					

Press [2nd] [STAT] [F2] and you will be on the statistical list-editor screen. Move the cursor to the top row and then to the right until you see a screen like the one on the next page. We are going to name the lists which we will use to work the problem.

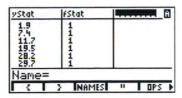

We are going to name the list SIZE. S, I, Z, and E are the alphabetic characters associated with the ⑥ ⑴ ⑴ ⌃ keys, respectively. We can do this since the calculator is in ALPHABETIC mode as indicated by the ⓐ in the upper right corner or the screen. Press [ENTER] and the list is named.

Press ⮟ to get into the list. Now enter 2500, 1250, 500, 250, 167, 125, 100, 90, and 50. After entering each value, press [ENTER] or ⮟ to drop down to the next line and make the next entry in the list. This list is too long to be seen on a single screen. Use the ⮝ and ⮟ keys to scroll up and down in the list and view all the data.

Once this is done the spread sheet ability of the calculator can be used to fill in the rest of the table. With the data in, move the cursor up to highlight "SIZE" and press ▶ to name the next list. Notice the ⓐ in the upper right corner. The calculator is in ALPHABETIC mode so we can name the list NUMBER by pressing the appropriate keys to spell the name of the list followed by [ENTER].

Use the right-arrow key and name the next list INVENT, for inventory. Continue moving to the right and naming the lists, CARYCOST, for carrying costs; ORDRCOST, for ordering costs; REORCOST, for reordering costs; and TOTLCOST, for total costs.

INVENT	CARYCOST	ORDRCOST 8
--------	--------	--------

ORDRCOST =

| { | } | NAMES | " | OPS ▶ |

CARYCOST	ORDRCOST	REORCOST 9
--------	--------	--------

REORCOST =

| { | } | NAMES | " | OPS ▶ |

ORDRCOST	REORCOST	TOTLCOST 10
--------	--------	--------

TOTLCOST =

| { | } | NAMES | " | OPS ▶ |

With all the lists named, use the left-arrow to move back to the list named NUMBER. With the name highlighted, press [2] [5] [0] [0] [÷] [F3] to see the listing of named lists at the bottom of the page. If "SIZE" is not displayed, press [MORE] until you find it. Then press the appropriate F key to choose it. Press [ENTER] and the list fills in. Since decimal parts of recorders makes no sense, the decimals in "NUMBER" can be replaced with whole numbers, if you wish.

SIZE	NUMBER	INVENT 5
2500	--------	--------
1250		
500		
250		
167		
125		

NUMBER =

| { | } | NAMES | " | OPS ▶ |

SIZE	NUMBER	INVENT
2500	--------	--------
1250		
500		
250		

NUMBER =2500/SIZE

| { | } | NAMES | " | OPS |
| SIZE | TOTLC | fStat | xStat | yStat |

SIZE	NUMBER	INVENT 5
2500	1	--------
1250	2	
500	5	
250	10	

NUMBER(1) =1

| { | } | NAMES | " | OPS |
| SIZE | TOTLC | fStat | xStat | yStat |

Move the cursor so that the name "INVENT" is highlighted. Then press the appropriate F key to choose "SIZE" and press [÷] [2] [ENTER] and the list will fill in.

SIZE	NUMBER	INVENT 6
2500	1	--------
1250	2	
500	5	
250	10	

INVENT =

| { | } | NAMES | " | OPS |
| SIZE | TOTLC | fStat | xStat | yStat |

SIZE	NUMBER	INVENT 6
2500	1	--------
1250	2	
500	5	
250	10	

INVENT =SIZE/2

| { | } | NAMES | " | OPS |
| SIZE | TOTLC | fStat | xStat | yStat |

SIZE	NUMBER	INVENT 6
2500	1	1250
1250	2	625
500	5	250
250	10	125

INVENT(1) =1250

| { | } | NAMES | " | OPS |
| SIZE | TOTLC | fStat | xStat | yStat |

Move the cursor so that the name "CARYCOST" is highlighted. Then press [1] [0] [×] and the appropriate F key to choose "INVENT" and press [ENTER] and the list will fill in. Remember that you many have to press [MORE] to find the "INVENT" in the list of names of the lists.

NUMBER	INVENT	CARYCOST 7
1	1250	--------
2	625	
5	250	
10	125	

CARYCOST =

| { | } | NAMES | " | OPS |
| SIZE | TOTLC | fStat | xStat | yStat |

NUMBER	INVENT	CARYCOST 7
1	1250	--------
2	625	
5	250	
10	125	

CARYCOST =10*INVENT

| { | } | NAMES | " | OPS |
| CARYC | INVEN | NUMB | ORDRC | REORC ▶ |

NUMBER	INVENT	CARYCOST 7
1	1250	12500
2	625	6250
5	250	2500
10	125	1250

CARYCOST(1) =12500

| { | } | NAMES | " | OPS |
| CARYC | INVEN | NUMB | ORDRC | REORC ▶ |

Move the cursor so that the name "ORDRCOST" is highlighted. Then press [2] [0] [+] [9] [×] and the appropriate F key to choose "SIZE" and press [ENTER] and the list will fill in.

INVENT	CARYCOST	ORDRCOST 8
1250	12500	--------
625	6250	
250	2500	
125	1250	

ORDRCOST =

| { | } | NAMES | " | OPS |
| CARYC | INVEN | NUMB | ORDRC | REORC ▶ |

INVENT	CARYCOST	ORDRCOST 8
1250	12500	--------
625	6250	
250	2500	
125	1250	

ORDRCOST =20+9*SIZE

| { | } | NAMES | " | OPS |
| SIZE | TOTLC | fStat | xStat | yStat |

INVENT	CARYCOST	ORDRCOST 8
1250	12500	22520
625	6250	11270
250	2500	4520
125	1250	2270

ORDRCOST(1) =22520

| { | } | NAMES | " | OPS |
| SIZE | TOTLC | fStat | xStat | yStat |

Move the cursor so that the name "REORCOST" is highlighted. Then press the appropriate F key to choose "ORDRCOST" and press [×]. Then press the appropriate F key to choose "NUMBER" and press [ENTER] to see the list fill in.

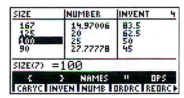

Move the cursor so that the name "TOTLCOST" is highlighted. Then press the appropriate F key to choose "CARYCOST" and press ⊞. Then press the appropriate F key to choose "REORCOST" and press ENTER to see the list fill in.

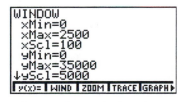

We search the "TOTLCOST" list for the smallest number and find that $23,500 is the lowest total inventory cost. We highlight that element and use the ◄ to return to SIZE. Here we find that a lot size of 100 recorders results in the lowest total inventory cost.

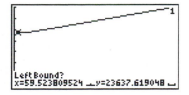 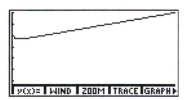

Use the values to fill in the table.

The following is an explanation of how to work the problem graphically.

1.   Graph $C(x) = 10 \cdot \dfrac{x}{2} + (20 + 9x)\dfrac{2500}{x}$ over the interval [0, 2500].

On the equation-editor screen, delete all previous entries and turn off the plots. Then enter the cost equation as $y1 = 10 \cdot \dfrac{x}{2} + (20 + 9x)\dfrac{2500}{x}$. Below is an appropriate window. Press GRAPH F5 after setting the window.

2.   Graphically estimate the minimum value and where it occurs.

Press GRAPH MORE F1 F4 to select "FMIN" from the GRAPH MATH menu. This feature has been used several times before in this manual. Select a left bound. Select a right bound. Make a guess.

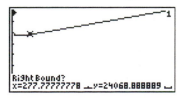 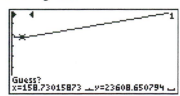

Once the guess has been made, the minimum appears on the screen.

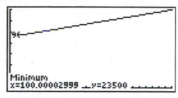

As we found in the table, the minimum total inventory cost is $23,500 and that occurs when the lot size is 100 television sets.

MARGINAL COST, REVENUE, AND PROFIT

Section 2.6, Example 1, page 276 *Business: Marginal Cost, Revenue, and Profit.* Given $C(x) = 62x^2 + 27,500$ and $R(x) = x^3 - 12x^2 + 40x + 10$ find each of the following:

(a)   total profit, $P(x)$.

Profit = Revenue – Cost

$$P(x) = x^3 - 12x^2 + 40x + 10 - (62x^2 + 27500)$$
$$P(x) = x^3 - 12x^2 + 40x + 10 - 62x^2 - 27500$$
$$P(x) = x^3 - 74x^2 + 40x - 27490$$

(b)   total cost, revenue, and profit from the production and sale of 50 units of the product.

On the equation-editor screen enter the cost function as y1 $= 62x^2 + 27,500$ and the revenue function as y2 $= x^3 - 12x^2 + 40x + 10$. Move the cursor immediately to the right of "y3=" and press [F2] [2] [–] [F2] [1]. Now we have the cost function in the calculator as y1, the revenue function as y2, and the profit function as y3.

To evaluate each of the three functions when 50 units are produced and sold, go to the home screen by pressing [2nd] [QUIT]. To find $C(50)$ press [2nd] [alpha] Y [1] [(] [5] [0] [)] [ENTER] and we see $C(50) = \$182,500$. To find $R(50)$ press [2nd] [alpha] Y [2] [(] [5] [0] [)] [ENTER] and we see $R(50) = \$97,010$. To find $P(50)$ press [2nd] [alpha] Y [3] [(] [5] [0] [)] [ENTER]   and we see $P(50) = -\$85,490$. Negative profit indicates a loss.

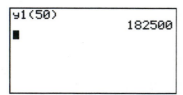

(c)   The marginal cost, revenue, and profit when 50 units are produced and sold. On the home screen, press [2nd] [CALC] [F2] to select "nDer" from the CALCULUS menu. To find $C'(50)$ press [2nd] [alpha] Y [1] [,] [x-VAR]

⌐,⌐ ⌐5⌐ ⌐0⌐ ⌐)⌐ ⌐ENTER⌐ and we see that $C'(50) = \$6200$. To find $R'(50)$ press ⌐F2⌐ ⌐2nd⌐ [alpha] Y ⌐2⌐ ⌐,⌐ ⌐x-VAR⌐ ⌐,⌐ ⌐5⌐ ⌐0⌐

⌐)⌐ ⌐ENTER⌐ and we see that $R'(50) = \$6340$. To find $P'(50)$ press ⌐F2⌐ ⌐2nd⌐ [alpha] Y ⌐3⌐ ⌐,⌐ ⌐x-VAR⌐ ⌐,⌐ ⌐5⌐ ⌐0⌐ ⌐)⌐ ⌐ENTER⌐

and we see that $P'(50) = \$140$.

```
nDer(y1,x,50)
                6200
■



evalF  nDer  der1  der2  fnInt ▶
```

```
nDer(y1,x,50)
                6200
nDer(y2,x,50)
                6340
■


evalF  nDer  der1  der2  fnInt ▶
```

```
nDer(y1,x,50)
                6200
nDer(y2,x,50)
                6340
nDer(y3,x,50)
                 140
■
evalF  nDer  der1  der2  fnInt ▶
```

Chapter 3
Exponential and Logarithmic Functions

GRAPHING LOGARITHMIC FUNCTIONS

Section 3.2, Technology Connection, page 320   Here we show two ways to graph $y = \log_2 x$.

First, we use the DRAW INVERSE, "DrInv," feature from the GRAPH DRAW menu. The equation, $y = \log_2 x$, must first be rewritten as the exponential equation $2^y = x$. These equations are different representations of the same equation. On the graphing calculator, we must have an equation solved for y in order to graph it. Therefore, we interchange the x and y in $2^y = x$ and graph $y1 = 2^x$. This equation is the inverse of $y = \log_2 x$. With $y1 = 2^x$ on the equation-editor screen, press GRAPH F3 F4 to graph the function on a standard viewing window. Now press GRAPH MORE F2 to reach the GRAPH DRAW menu. Now press MORE MORE MORE and notice the bottom of the screen. Choose "DrInv" by pressing F3 and the command "DrInv" is pasted on the home screen. Press 2nd [alpha] Y 1.

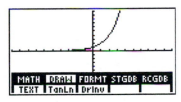

Press ENTER again and the graph and its inverse are displayed on the standard viewing window. The graph of the inverse of y1 is the graph of $y = \log_2 x$.

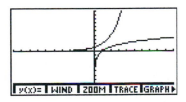

The second way to graph $y = \log_2 x$ is to use the Change-of-Base Formula. See Theorem III on page 321 in your textbook. The Change-of-Base Formula is written here for clarity: $\log_b M = \dfrac{\log_a M}{\log_a b}$. There are two logarithm buttons on the calculator. One is the LOG key. It represents $\log_{10}$ which is referred to as a common logarithm. The other is the LN key. It represents $\log_e$ and is referred to as a natural logarithm. Either can be used in the Change-of-Base Formula. The equation $y = \log_2 x$ can be rewritten as $y = \dfrac{\log (x)}{\log (2)}$ or $y = \dfrac{\ln (x)}{\ln (2)}$ using the Change-of-Base Formula.

On the equation-editor screen clear all entries and turn off all plots. Enter $y1 = \dfrac{\log (x)}{\log (2)}$ and press GRAPH F3 F4 to graph the function on the standard viewing window.

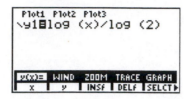 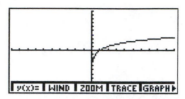

With this graph the TABLE is available and so is the TRACE feature. Using DRAW INVERSE, "DrInv," none of the regular features associated with a graph are accessible.

SOLVING EXPONENTIAL EQUATIONS

Section 3.2, Technology Connection, page 325   Solve $e^t = 40$ using a graphing calculator.

Either the INTERSECT feature or the ROOT feature could be used. Here we use the INTERSECT feature. The ROOT feature is discussed on page 97 in this manual. Put the left side of the equation in y1 and the right side in y2 on the equation-editor screen. With the cursor immediately to the right of "y1=" press [2nd] [e^x] [x-VAR]. ([e^x] is the second function associated with the [LN] key.) With the cursor immediately to the right of "y2=" press [4] [0]. There are many choices for a good viewing window, but the maximum y-value must be at least 40 since we will have a horizontal line at $y = 40$. Once the window is set, press [GRAPH] [F5].

 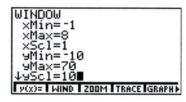 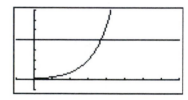

With the graph displayed, press [GRAPH] [MORE] [F1] [MORE] [F3] to choose "ISECT." The calculator will pose three questions.

The query "First Curve?" appears at the bottom of the screen. The blinking cursor is positioned on the graph of y1 as in the screen below on the left. Notice the upper right corner of this screen displays 1, indicating that this is the equation in y1 on the equation-editor screen. Press [ENTER] to indicate that this is the first curve involved in the intersection.

Next the query "Second Curve?" appears at the bottom of the screen. The blinking cursor is now positioned on the graph of y2 as in the screen below on the right. Notice the upper right corner of this screen displays 2, indicating that this is the equation in y2 on the equation-editor screen. Press [ENTER] to indicate that this is the second curve.

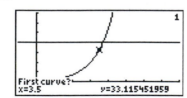 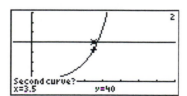

After we identify the second curve, the query "Guess?" appears at the bottom of the screen. Since there is only one point of intersection, press [ENTER] and the point of intersection is displayed.

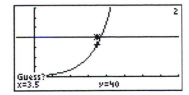 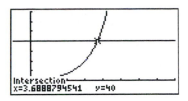

The *x*-coordinate is the solution to the equation. The solution is $t \approx 3.7$.

EXPONENTIAL REGRESSION

Section 3.3, Technology Connection, page 341 The table below shows data regarding world population growth.

Year	World Population (in billions)
1927	2
1960	3
1974	4
1987	5
1998	6

To find an exponential equation that models the data, enter the data as ordered pairs on the statistical list-editor screen. Press [2nd] [STAT] [F2]. To clear the existing entries in the lists, use the arrow keys to move the cursor to highlight "xStat." Press [CLEAR] [ENTER]. Now highlight "yStat." Press [CLEAR] [ENTER].

Once the lists are cleared, we can enter the data points. We will enter the number of years in xStat and the population in yStat. Position the cursor just below "xStat." Press [1] [9] [2] [7] [ENTER]. Continue typing the *x*-values, 1960, 1974, 1987, and 1998, each followed by [ENTER]. The entries can be followed by [▼] rather than [ENTER] if desired. Press [▶] to move the cursor just below "yStat." Type the population in billions 2, 3, 4, 5, and 6, each followed by [ENTER] or [▼]. Note that the coordinates of each point must be in the same position in both lists.

```
xStat     yStat     fStat      2
1927      2         1
1960      3         1
1974      4         1
1987      5         1
1998      6         1
--------            1
yStat(6) =
{    }    NAMES    "    OPS  ▶
```

The calculator's EXPONENTIAL REGRESSION feature can be used to fit an exponential equation to the data. With the data points in the lists, press [2nd] [QUIT] to go get out of the editor and go to the home screen. Press [2nd] [STAT] [F1] and look at the bottom of the screen. Press [F5] to choose "ExpR" from the STATISTICAL CALCULATIONS menu. Now press [2nd] [LIST] [F3] and choose "xStat" by pressing the appropriate F key. If you do not see the name "xStat," press the [MORE] key until you do. Then press [,] and the appropriate F key to select "yStat." If you do not see the name "yStat" press the [MORE] key until you do. This is done to let the calculator know which lists contain the data. (Even though the calculator was programmed to assume the *x*- and *y*-coordinates of data points are in "xStat" and "yStat," respectively, it is good practice to specify the lists being used.) Before the regression equation is found, it is possible to select a *y*-variable to which the equation will be stored on the equation-editor screen. To do this, press ⌐,⌐ ⌐2nd⌐ [alpha] Y ⌐1⌐. Finally, to display the coefficients of the regression equation

press ⌐ENTER⌐. Press ⌐EXIT⌐ to remove the menus from the screen if you wish.

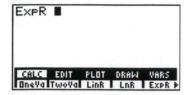

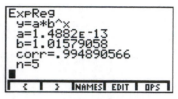

With the equation in y1, we evaluate y1(2008). This should predict the world's population in 2008. On the

home screen, press ⌐2nd⌐ [alpha] Y ⌐1⌐ ⌐(⌐ ⌐2⌐ ⌐0⌐ ⌐0⌐ ⌐8⌐ ⌐)⌐ ⌐ENTER⌐.

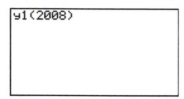
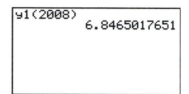

In 2008, the world's population will be approximately 6.8465 billion.

EXPONENTIAL DECAY

Section 3.4, Example 1, page 353 *Life Science: Decay* Strontium-90 has a decay rate of 2.8% per year. The rate

of change of an amount N is given by $\dfrac{dN}{dt} = -0.028N$.

(a)   Find the function that satisfies the equation. Let N_0 represent the amount present at $t = 0$.

Since $\dfrac{dP}{dt} = -kP$ and we have $\dfrac{dN}{dt} = -0.028N$, we know $k = 0.028$. Using the function $P(t) = P_0 e^{-kt}$ and

substituting, we have $N(t) = N_0 e^{-0.028t}$.

(b)   Suppose that 1000 grams of strontium-90 is present at $t = 0$. How much will remain after 70

years?

The function in part (a) is now $N(t) = 1000 e^{-0.028t}$. On the equation-editor screen enter y1 $= 1000 e^{-0.028x}$.

The t in the function is x in the calculator. Remember $[e^x]$ is the second function associated with the ⌐LN⌐ key. Press

⌐2nd⌐ [QUIT] to go to the home screen. To evaluate the function when $t = 70$ press ⌐2nd⌐ [alpha] Y ⌐1⌐ ⌐(⌐ ⌐7⌐ ⌐0⌐ ⌐)⌐ ⌐ENTER⌐.

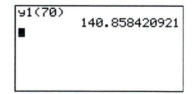

After 70 years, about 140.9 grams of strontium-90 remains.

(c)   After how long will half of the 1000 grams remain? (Here we are finding the half-life of the

substance.) With the function already in y1, use y2 = 500 as the second equation. The x-coordinate of the

intersection point will be the answer to the question.

The *x*-axis is the time axis and therefore should not have any negative values. The *y*-axis is the axis which gives the quantity of the substance, so it should not be negative. The answer is in first quadrant. Trial and error will be necessary to find a good window. The one below works well. Once we have the desired window, press GRAPH F5 to see the function and the horizontal line displayed.

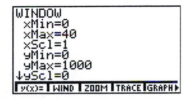
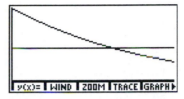

With the graph displayed, press GRAPH MORE F1 MORE F3 to choose "ISECT". The calculator will pose three questions.

The query "First Curve?" appears at the bottom of the screen. The blinking cursor is positioned on the graph of y1 as in the screen below on the left. Press ENTER to indicate that this is the first curve involved in the intersection.

Next the query "Second Curve?" appears at the bottom of the screen. The blinking cursor is now positioned on the graph of y2 as in the screen below on the right. Press ENTER to indicate that this is the second curve.

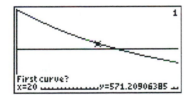
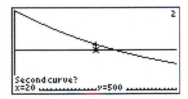

After we identify the second curve, the query "Guess?" appears at the bottom of the screen. Since there is only one point of intersection, press ENTER and the point of intersection is displayed.

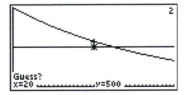
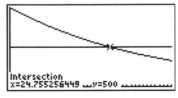

The *x*-coordinate is the solution to the equation, $t \approx 24.755$ years. We can write this answer in years, months, and days by immediately going to the home screen. Press 2nd [QUIT] to do this. Once on the home screen press x-VAR ENTER to see the answer revealed. Subtract the years away by pressing − 2 4 ENTER. Notice that "Ans" is displayed when − is pressed.

```
┌─────────────────────────────┐   ┌─────────────────────────────┐
│ x                           │   │ x                           │
│            24.7552564486    │   │            24.7552564486    │
│                             │   │ Ans-24                      │
│                             │   │             .75525644857    │
│                             │   │                             │
└─────────────────────────────┘   └─────────────────────────────┘
```

The number which remains is the decimal part of a year. To convert it to months, multiply by 12 by pressing ⊠ ① ② ENTER. "Ans" is displayed again. We see that the decimal converted to 9 months.

```
┌─────────────────────────────┐
│ x                           │
│            24.7552564486    │
│ Ans-24                      │
│             .75525644857    │
│ Ans*12                      │
│             9.06307738284   │
│ ■                           │
└─────────────────────────────┘
```

Subtract away the months by pressing ⊟ ⑨ ENTER. The decimal which remains is the decimal part of a month. To convert this to days we need to multiply it by the number of days in a month. Since the number of days in a month varies, use 30 as a close approximation. Multiply by 30 by pressing ⊠ ③ ⓪ ENTER. The decimal converts to approximately 2 days.

```
┌─────────────────────────────┐   ┌─────────────────────────────┐
│            24.7552564486    │   │             .75525644857    │
│ Ans-24                      │   │ Ans*12                      │
│             .75525644857    │   │             9.06307738284   │
│ Ans*12                      │   │ Ans-9                       │
│             9.06307738284   │   │             .06307738284    │
│ Ans-9                       │   │ Ans*30                      │
│             .06307738284    │   │             1.8923214852    │
│ ■                           │   │ ■                           │
└─────────────────────────────┘   └─────────────────────────────┘
```

The half-life of the substance is 24 years, 9 months, and 2 days.

PRESENT VALUE

Section 3.4, Example 4, page 356 *Business: Present Value* Following the birth of their grand daughter, two grandparents want to make an initial investment of P_0 that will grow to \$10,000 by the child's 20th birthday. Interest is compounded continuously at 6%. What should the initial investment be?

Using the formula for continuous compounding, $P = P_0 e^{kt}$, we have $10,000 = P_0 e^{0.06 \cdot 20}$. This can be solved using the INTERSECT feature of the GRAPH MATH menu by putting the left side of the equation in y1 and the right side in y2. P_0 , the present value, is x in the calculator. Remember $[e^x]$ is the second function associated with the LN key. The 6% interest goes into the formula where the "k" was. Remember $6\% = .06$. The "t" in the formula is time and it is 20 years.

The x-axis represents the present value and the y-axis is the future value on the calculator. Negative values would make no sense for either. This means we need to only look in first quadrant. Trial and error will be necessary to find a good window. The one below works well. Once we have the desired window, press [GRAPH] [F5] to see the function and the horizontal line displayed. Press [CLEAR] to remove the menu from the bottom of the screen.

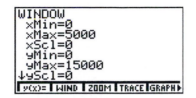
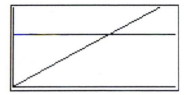

Press [GRAPH] [MORE] [F1] [MORE] [F3] to choose "ISECT." The calculator will pose three questions.

The query "First Curve?" appears at the bottom of the screen. The blinking cursor is positioned on the graph of y1 as in the screen below on the left. Press [ENTER] to indicate that this is the first curve involved in the intersection.

Next the query "Second Curve?" appears at the bottom of the screen. The blinking cursor is now positioned on the graph of y2 as in the screen below on the right. Press [ENTER] to indicate that this is the second curve.

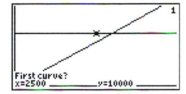
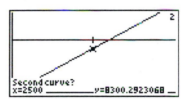

After we identify the second curve, the query "Guess?" appears at the bottom of the screen. Since there is only one point of intersection, press [ENTER] and the point of intersection is displayed.

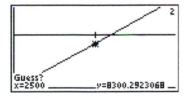
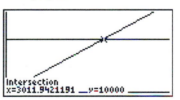

The x-coordinate is the solution to the equation. The grandparents need to deposit $3011.94 at the child's birth for it to grow to $10,000 by the child's 20th birthday.

Chapter 4
Integration

SUMMATION NOTATION, $\sum$

Summations can be evaluated using the SUMMATION and SEQUENCE features from the LIST OPERATIONS menu on the graphing calculator.

Section 4.1, Example 5, page 395 Express $\sum_{i=1}^{4} 3^i$ without using summation notation. Here we evaluate $\sum_{i=1}^{4} 3^i$ using the graphing calculator. The commands are "sum seq(" followed by the sequence, the variable name, the beginning number, the ending number, and how to increment. Then the parentheses are closed. Press [2nd] [LIST] [F5] [MORE]. Look at the menu across the bottom of the screen. Press [F1] to choose "sum" and then press [F3] to choose "seq(."

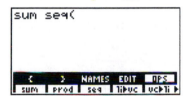

Press [3] [^] [ALPHA] I. (The letter I is the alphabetic character associated with the [)] key.) With the sequence defined, press [,] [ALPHA] I to name the variable.

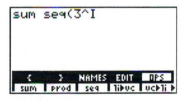

Press [,] [1] [,] [4]. This tells the calculator where to start and where to stop. Press [,] [1] [)] to tell the calculator how to count and to close the parentheses. Press [ENTER] and the answer is revealed. $\sum_{i=1}^{4} 3^i = 120$.

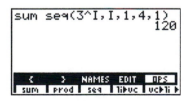

THE fnInt FEATURE

Definite integrals can be evaluated on the home screen using the NUMERICAL FUNCTION INTEGRAL feature from the CALCULUS menu.

Section 4.3, Technology Connection, page 418 Evaluate $\int_{-1}^{2}(-x^3+3x-1)\,dx$ using the NUMERICAL FUNCTION

INTEGRAL feature of the graphing calculator.

First select "fnInt(" from the CALCULUS menu by pressing [2nd] [CALC] [F5]. Then enter the function, the variable, and the lower and upper limits of integration. Press [(-)] [x-VAR] [^] [3] [+] [3] [x-VAR] [-] [1] [,] [x-VAR] [,] [(-)] [1]

[,] [2] [)] [ENTER]. We find that $\int_{-1}^{2}(-x^3+3x-1)\,dx = -2.25$.

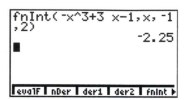

If the function has been entered on the equation-editor screen, as $y1 = -x^3+3x-1$ we can evaluate it by entering fnInt(y1, x, -1, 2) on the home screen. Press [2nd] [QUIT] [2nd] [CALC] [F5] [2nd] [alpha] Y [1] [,] [x-VAR] [,] [(-)] [1]

[,] [2] [)] [ENTER].

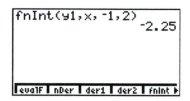

THE $\int f(x)\,dx$ FEATURE

Definite integrals can also be evaluated on the graph screen using the NUMERICAL INTEGRAL feature,

" $\int f(x)\,dx$," from the GRAPH MATH menu.

Section 4.3, Technology Connection, page 418 Evaluate $\int_{-1}^{2}(-x^3+3x-1)\,dx$ on the graph screen, using the

NUMERICAL INTEGRAL feature from the GRAPH MATH menu.

First, graph $y1 = -x^3+3x-1$ in a window that contains the interval [-1, 2]. We will use [-3, 3, -6, 6]. Then select the " $\int f(x)\,dx$ " feature from the GRAPH MATH menu by pressing [GRAPH] [MORE] [F1] [F3].

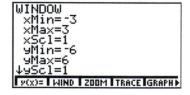
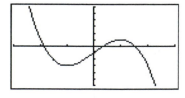

We are prompted to enter the lower limit of integration. Press [(-)] [1]. Then press [ENTER]. Now enter the upper limit of integration by pressing [2]. Press [ENTER] again. The calculator shades the area above and below the curve on [-1, 2] and returns the value of the definite integral on this interval.

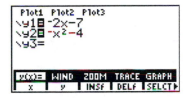 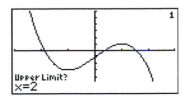 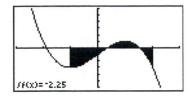

THE fnInt FEATURE; AREA BETWEEN TWO CURVES

The area between two curves can be found using the INTERSECT feature from the GRAPH MATH menu and "fnInt(" from the CALCULUS menu.

Section 4.4, Technology Connection, page 429 Find the area bounded by the two graphs $y1 = -2x - 7$ and $y2 = -x^2 - 4$. First graph both functions and determine which function is the upper and which is the lower graph. The equation-editor screen, the window, and the graph are displayed below. Remember to press $\boxed{\text{CLEAR}}$ to remove the menu from the graph.

 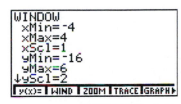 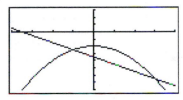

To determine which curve is the upper one, press $\boxed{\text{GRAPH}}$ $\boxed{\text{F4}}$. A flashing cursor will appear on y1 and 1 will be displayed in the upper right corner of the screen. Press either the up- or down-arrow key to move to y2 and 2 will be displayed. From examining these screens we see that y2 is the upper curve for the region between the two intersection points.

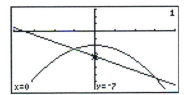 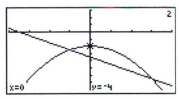

Go to the home screen by pressing $\boxed{\text{2nd}}$ [QUIT]. Select "fnInt(" by pressing $\boxed{\text{2nd}}$ [CALC] $\boxed{\text{F5}}$. To enter the function, enter the upper curve, y2, minus the lower curve, y1. Then enter the variable. Press $\boxed{\text{2nd}}$ [alpha] Y $\boxed{2}$ $\boxed{-}$ $\boxed{\text{2nd}}$ [alpha] Y $\boxed{1}$ $\boxed{,}$ $\boxed{\text{x-VAR}}$ $\boxed{,}$. To find the lower and upper limits of integration, return to the graph screen by pressing $\boxed{\text{GRAPH}}$.

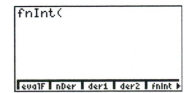 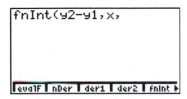

Use the INTERSECT feature from the GRAPH MATH menu to find the points of intersection. The x-coordinates of the points are the lower and upper limits of integration. With the graph displayed, press $\boxed{\text{MORE}}$ $\boxed{\text{F1}}$ $\boxed{\text{MORE}}$ $\boxed{\text{F3}}$ to choose "ISECT."

The calculator will pose three questions.

The query "First Curve?" appears at the bottom of the screen. The blinking cursor is positioned on the graph of y1 as in the screen below on the left. Press [ENTER] to indicate that this is the first curve involved in the intersection.

Next the query "Second Curve?" appears at the bottom of the screen. The blinking cursor is now positioned on the graph of y2 as in the screen below on the right. Press [ENTER] to indicate that this is the second curve.

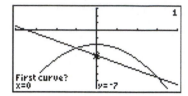
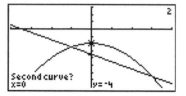

After we identify the second curve, the query "Guess?" appears at the bottom of the screen. Press [ENTER] and the point of intersection nearest the cursor is displayed.

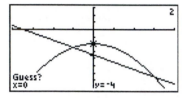
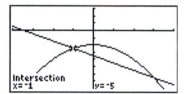

This *x*-coordinate is the lower limit of integration. So the lower limit of integration is −1.

We repeat the process to find the upper limit of integration, which is the other point of intersection. Press [GRAPH] [MORE] [F1] [MORE] [F3] to choose "ISECT." The calculator will pose three questions.

The query "First Curve?" appears at the bottom of the screen. The blinking cursor is positioned on the graph of y1 as in the screen below on the left. Press [ENTER] to indicate that this is the first curve involved in the intersection.

Next the query "Second Curve?" appears at the bottom of the screen. The blinking cursor is now positioned on the graph of y2 as in the screen below on the right. Press [ENTER] to indicate that this is the second curve.

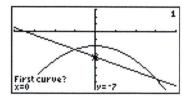
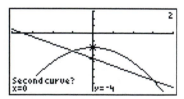

After we identify the second curve, the query "Guess?" appears at the bottom of the screen. This time move the cursor near the point of intersection in the fourth quadrant, since that is the one we have not found. When the cursor is near it, press [ENTER].

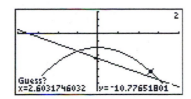 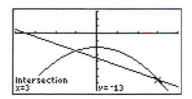

This x-coordinate is the upper limit of integration. So the upper limit of integration is 3. Press [2nd] [QUIT] to return to the home screen.

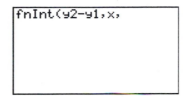

The calculator should be waiting for the lower and upper limits of integration. Press [(-)] [1] [,] [3] [)]. Then press [ENTER]. To express the answer as a fraction, press [2nd] [MATH] [F5] [MORE] [F1] [ENTER].

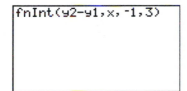 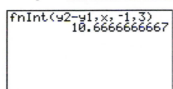 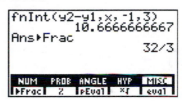

The area between the two curves is $10\frac{2}{3}$ square units or $\frac{32}{3}$ square units.

AVERAGE VALUE

Section 4.4, Example 5, page 432 Find the average value of $f(x) = x^2$ over the interval [0, 2].

The formula to find the average value of a function over a closed interval is given by: $y_{av} = \frac{1}{b-a}\int_a^b f(x)$.

For this example $a = 0$ and $b = 2$. Go to the home screen and press [(] [1] [÷] [(] [2] [-] [0] [)] [)] [2nd] [CALC] [F5] [x-VAR] [x²] [,] [x-VAR] [,] [0] [,] [2] [)]. Press [ENTER] to find the average value over the interval. To convert the answer into a fraction, press [2nd] [MATH] [F5] [MORE] [F1] [ENTER].

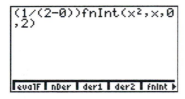 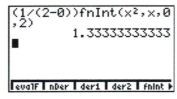 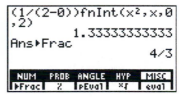

The average value of $f(x) = x^2$ over the interval [0, 2] is $\frac{4}{3}$.

Chapter 5
Applications of Integration

AN ECONOMICS APPLICATION: CONSUMER and PRODUCER SURPLUS

Section 5.1, Example 3, page 474   Given $D(x) = (x-5)^2$ and $S(x) = x^2 + x + 3$,

    (a)   find the equilibrium point.

To find the equilibrium point, graph both the supply and demand curves in an appropriate viewing window.

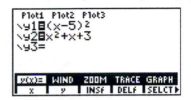 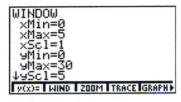 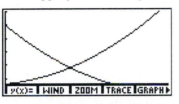

Use "ISECT" from the GRAPH MATH menu to find the equilibrium point. Press GRAPH MORE F1 MORE F3 to choose "ISECT" and respond to each of the three questions "First Curve?," "Second Curve?," and "Guess?" by pressing ENTER each time.

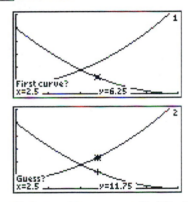 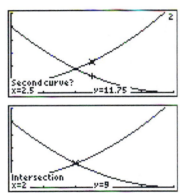

The equilibrium point occurs at (2, $9). The equilibrium quantity is 2 items and the equilibrium price is $9.

    (b)   find the consumer surplus at the equilibrium point.

The region between the two curves, the y-axis, and the equilibrium point contains the area that represents both the producer surplus and consumer surplus. First separate the region into two parts by graphing y3 = 9. This will put a horizontal line through the equilibrium point.

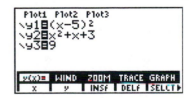 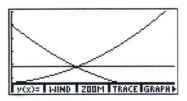

The upper region represents the consumer surplus as is illustrated on the screen on the next page. This screen is provided to clarify which region's area we are finding. Instructions to shade the region are not given here.

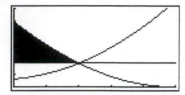

To find the upper and lower curves before integrating, press GRAPH F4 then use the ◄ key to move the cursor to part of the curve which serves as a boundary of the region in question. We use the ▲ and ▼ keys to move from curve to curve. We see that y1 is the upper boundary and y3 is the lower boundary of this region. The numbers 1, 2, or 3 are displayed in the upper right corner of the screen, indicating which curve the cursor is on.

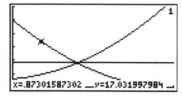 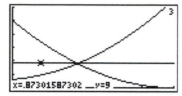 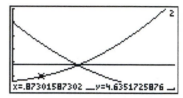

To find the consumer surplus, go to the home screen by pressing 2nd [QUIT]. We will evaluate the definite integral on the home screen using the NUMERICAL FUNCTION INTEGRAL feature from the CALCULUS menu. Press 2nd [CALC] F5 to select "fnInt(." Since y1 is the upper boundary of the region and y3 is the lower, press 2nd [alpha] Y 1 − 2nd [alpha] Y 3 , x-VAR , to enter the upper boundary minus the lower boundary and the variable name. To integrate from the y-axis to the equilibrium point, press 0 , 2). Then press ENTER. The consumer surplus is $14.67.

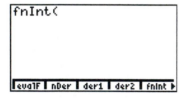 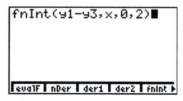 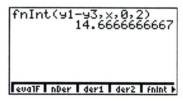

(c)   find the producer surplus at the equilibrium point.

The region between the two curves, the y-axis, and the equilibrium point contains the area that represents both the producer surplus and consumer surplus. We have separated the region into two parts by graphing y3 = 9.

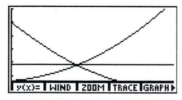

The lower region represents the producer surplus as is illustrated on the screen below. This screen is provided to clarify which area we are finding. Instructions to shade the region are not given here.

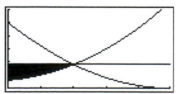

To find the upper and lower curves before integrating, press GRAPH F4 then use the ◄ key to move the cursor to part of the curve which serves as a boundary of the region in question. We use the ▲ and ▼ keys to move from curve to curve. We see that y3 is the upper boundary and y2 is the lower boundary of this region. The numbers 1, 2, or 3 are displayed in the upper right corner of the screen, indicating which curve the cursor is on.

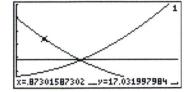 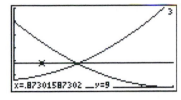 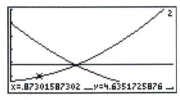

To find the producer surplus, go to the home screen by pressing 2nd [QUIT]. We will use the NUMERICAL FUNCTION INTEGRAL, "fnInt(," feature from the CALCULUS menu. Press 2nd [CALC] F5 to select "fnInt(." Since y3 is the upper boundary of the region and y2 is the lower, press 2nd [alpha] Y 3 – 2nd [alpha] Y 2 , x-VAR , to enter the upper boundary minus the lower boundary and the variable name. To integrate from the y-axis to the equilibrium point, press 0 , 2). Then press ENTER. The producer surplus is $7.33.

 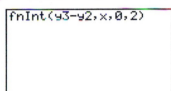 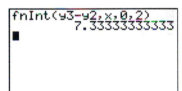

ACCUMULATED PRESENT VALUE

Section 5.2, Example 7, page 482 *Business: Accumulated Present Value* Find the accumulated present value of an investment over a 5-year period if there is a continuous money flow of $2400 per year and the interest rate is 14%, compounded continuously.

The accumulated present value is $\int_0^5 \$2400e^{-0.14t}\,dt$. This definite integral can be evaluated on the home screen using the NUMERICAL FUNCTION INTEGRAL, "fnInt(," feature from the CALCULUS menu. On the home screen, press 2nd [CALC] F5 2 4 0 0 2nd [e^x] ((-) . 1 4 2nd [alpha] T) , 2nd [alpha] T , 0 , 5). ([e^x] is the second function associated with the LN key and T is the alphabetic character associated with the – key. We get a lower case T by first pressing the 2nd key.) Press ENTER to see that the accumulated present value is $8629.97.

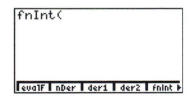 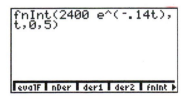 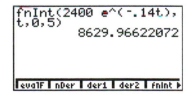

STATISTICS

Section 5.5, Example 4, page 511   The weights w of the students in a calculus class are normally distributed with a mean of 150 pounds and standard deviation of 25 pounds. Find the probability that a student's weight is from 160 to 180 pounds.

The TI-86 does not have the capability to graph a probability density function. Check the Texas Instruments web site—www.education.ti.com, for available programs to help with this process.

Chapter 6
Functions of Several Variables

PARTIAL DERIVATIVES

Section 6.2, Technology Connection, page 550   Given the function $f(x, y) = 3x^3 y + 2xy$, use a graphing

calculator that finds derivatives of functions of one variable to find $f_x(-4,1)$ and $f_y(2,6)$.

To find $f_x(-4,1)$, first find $f(x,1)$:

$$f(x, y) = 3x^3 y + 2xy$$
$$f(x,1) = 3x^3(1) + 2x(1)$$
$$= 3x^3 + 2x$$

Now we have a function with one variable. Both of the NUMERICAL DERIVATIVE features are used

below to solve the problem.

First we use "nDer(," from the CALCULUS menu to find the value of the derivative of this function

when $x = -4$. Enter the function on the equation-editor screen. Press [2nd] [QUIT] to go to the home screen. Press [2nd]

[CALC] [F2] to choose "nDer(" from the CALCULUS menu. Then press [2nd] [alpha] Y [1] [,] [x-VAR] [,] [(-)] [4] [)] [ENTER].

 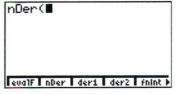 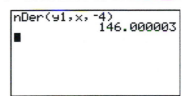

The procedures the calculator uses to calculate the derivative do not always yield an exact answer. Note

that the exact answer is 146, but the calculator produced 146.000003.

Now we use "dy/dx" from the GRAPH MATH menu to find the value of the derivative of this function

when $x = -4$. With the equation on the equation-editor screen, press [GRAPH] [F3] [F4] to see the graph displayed in a

standard viewing window. Press [GRAPH] [MORE] [F1] [F2] to choose "dy/dx" from the GRAPH MATH menu. Press

[(-)] [4] [ENTER].

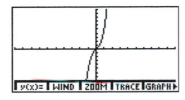 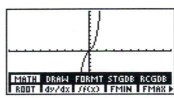

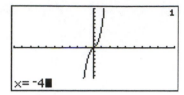 

This method yields the exact answer.

To find $f_y(2,6)$, first find $f(2,y)$:

$$f(x,y) = 3x^3 y + 2xy$$
$$f(2,y) = 3(2)^3 y + 2(2) y$$
$$= 24y + 4y$$
$$= 28y$$

Now find the derivative of $f(y) = 28y$ when $y = 6$ using "nDer(" from the CALCULUS menu by going to the home screen and pressing [2nd] [CALC] [F2] to choose "nDer(" from the CALCULUS menu. Press [2] [8] [ALPHA] Y [,] [ALPHA] Y [,] [6] [)] [ENTER]. (Y is the alphabetic character associated with the [0] key.)

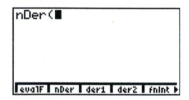 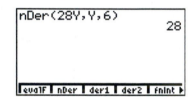

We can also replace y with x and find the derivative of $f(x) = 28x$ when $x = 6$ using "dy/dx" on the graph screen. Enter the function on the equation-editor screen. Change your window values to the ones below and press [GRAPH] [F5].

 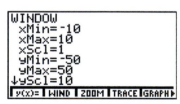 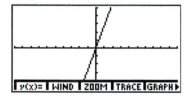

Press [GRAPH] [MORE] [F1] [F2] [6] [ENTER].

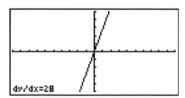

LINEAR REGRESSION

We can use the LINEAR REGRESSION feature in the STATISTICAL CALCULATIONS menu to fit a linear equation to a set of data points.

Section 6.4, page 566 Using the data in the table find the line of best fit. The least-squares technique is shown in the text.

The following table shows the yearly revenue for a rental car company that rents hybrid cars.

Years (in 5 year increments), x	1. 1991	2. 1996	3. 2001	4. 2006	5. 2011
Yearly Revenue (in millions), y	$5.2	$8.9	$11.7	$16.8	$

(a)   Make a scatter plot of the data and determine whether the data seem to fit a linear function.

We enter the data as ordered pairs on the statistical list-editor screen. Press 2nd [STAT] F2 and use the arrow keys to move the cursor to highlight "xStat." Press CLEAR ▼ or CLEAR ENTER to clear the list. Now highlight "yStat." Press CLEAR ▼ or CLEAR ENTER to clear the list. Your lists may contain different data, but whatever is there must be deleted in order to proceed.

Once the lists are cleared, we can enter the data points. We enter the number of years in xStat and the revenue amounts in yStat. Position the cursor directly below "xStat." Press 1 ENTER. Continue typing the x-values 2 through 4, each followed by ENTER. The entries can be followed by ▼ rather than ENTER if desired. Press ▶ to position the cursor directly below "yStat." Type the percentages 5.2, 8.9, and so on in succession, each followed by ENTER or ▼. Note that the coordinates of each point must be in the same position in both lists.

To plot the data points, we turn on the STATISTICAL PLOT feature. To access the statistical plot screen, press 2nd [STAT] F3.

We will use Plot 1. We access it by pressing F1. The cursor should be positioned over "On" and it should be flashing. Press ENTER to turn on Plot 1. The entries Type, Xlist, and Ylist should be as shown on the following page. The last item, Mark, allows us to choose a box, a cross, or a dot for each point. Here we have selected a box.

To select Type, Xlist Name, Ylist Name, and Mark, position the cursor to the right of whichever one you are choosing and look at the options at the bottom of the screen. Make the selection by pressing the appropriate F key.

        The plot can also be turned on from the equation-editor screen. Press [GRAPH] [F1]. Then assuming that Plot 1 has not yet been turned on and that the desired settings are currently entered for Plot 1 on the statistical plot screen, position the cursor over "Plot 1" and press [ENTER]. "Plot 1" will be highlighted. Before viewing the plot any existing entries on the equation-editor screen should be cleared. The easiest way to choose a viewing window is to use the ZOOM DATA feature. Press [GRAPH] [F3] [MORE] [F5] [CLEAR] to choose "ZDATA." "ZDATA" automatically defines a viewing window that displays all of the data points.

  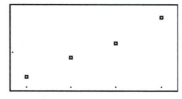

        With all the data points in the viewing window it is easy to modify the window to suit your needs. Below the window has been changed and the new graph is displayed.

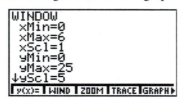 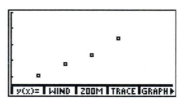

        The scatter plot shows that the data is approximately linear.

        (b)   Find a linear function that (approximately) fits the data.

        The calculator's LINEAR REGRESSION feature can be used to fit a linear equation to the data. With the data points in the lists, press [2nd] [STAT] [F1] [F3] to select "LinR" from the STATISTICAL CALCULATIONS menu. Now press [2nd] [LIST] [F3] and you see all the named lists in the calculator displayed on the bottom of the screen. Press the appropriate F key to choose "xStat" and "yStat." If you do not see them on the screen, press [MORE] until you do. Remember to separate them with [,]. We do this to let the calculator know which lists contain the data. (Even though the calculator was programmed to assume the x- and y-coordinates of data points are in xStat and yStat, respectively, it is good practice to specify the lists being used.) Before the regression equation is found, it is possible to select a y-variable to which it will be stored on the equation-editor screen. To do this, press [,] [2nd] [alpha] Y [1]. Finally, to display the coefficients of the regression equation press [ENTER] [EXIT]. The correlation coefficient is r. The correlation coefficient is used to describe the strength of the linear relationship between the data points. The closer r is to 1, the better the correlation.

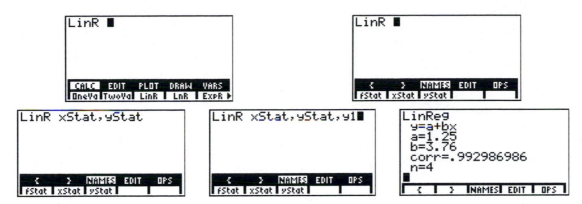

Press [GRAPH] [F5] to see the line displayed with the data points. The line of best fit is $y = 1.25 + 3.76x$.

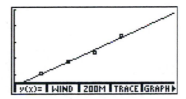

(c)   Use the model to predict the yearly revenue in 2011.

To predict the yearly revenue in 2011, evaluate the regression equation for $x = 5$. (2011 is 20 years after 1991 and in this example we assigned 1 to 1991 and are counting in 5 year increments.) Use any of the methods for evaluating a function presented earlier in this manual. We will use the FORECAST feature in the STATISTICAL menu. Press [2nd] [STAT] [MORE] [F1] to choose "FCST." Now the cursor is to the right of "x =." Press [5] [ENTER]. The cursor is to the right of "y =." Press [F5] to choose "SOLVE."

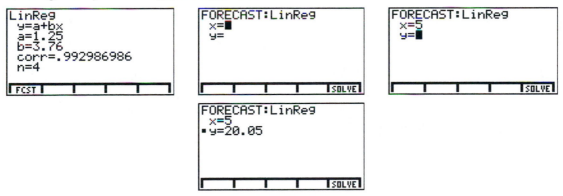

When $x = 5$, $y \approx 20.05$, so we predict 2011 yearly revenue from renting hybrid cars will be approximately $20.05 million.

USING 2-VARIABLE STATISTICS

We can use the 2-VARIABLE STATISTICS feature in the STATISTICAL CALCULATIONS menu to fit a linear equation to a set of data points.

Section 6.4, page 569   Using the data in the table find the line of best fit using the 2-VARIABLE STATISTICS feature, and the lists.

The table shows the yearly revenue for a rental car company that rents hybrid cars.

Years (in 5 year increments), x	1. 1991	2. 1996	3. 2001	4. 2006	5. 2011
Yearly Revenue (in millions), y	$5.2	$8.9	$11.7	$16.8	$

Press [2nd] [STAT] [F2] and enter the data into xStat and yStat.

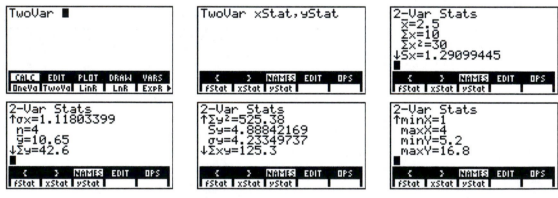

We wish to fill in the table and find the regression line.

Variable	c_i	d_i	$c_i - \overline{x}$	$(c_i - \overline{x})^2$	$d_i - \overline{y}$	$(c_i - \overline{x})(d_i - \overline{y})$
Calculator location	xStat	yStat	L3=xStat$-\overline{x}$	L4= L3^2	L5 = yStat$-\overline{y}$	L6 = L3 • L5
	1	5.2				
	2	8.9				
	3	11.7				
	4	16.8				
	$\sum_{i=1}^{4} c_i =$	$\sum_{i=1}^{4} d_i =$		$\sum_{i=1}^{4}(c_i - \overline{x})^2 =$		$\sum_{i=1}^{4}(c_i - \overline{x})(d_i - \overline{y}) =$
	$\overline{x} =$	$\overline{y} =$				

Press [2nd] [QUIT] [2nd] [STAT] [F1] [F2] to choose "TwoVar." Now press [2nd] [LIST] [F3] and press the appropriate F key to choose "xStat" and then "yStat" separated by a [,] to tell the calculator where the data is located. Press [ENTER] to reveal the 2-VARIABLE STATISTICS. Use the up- and down-arrow keys to scroll through the screen.

So far we have $\overline{x} = 2.5$, $\overline{y} = 10.65$, $\sum_{i=1}^{4} c_i = 10$, and $\sum_{i=1}^{4} d_i = 42.6$. To find $c_i - \overline{x}$ return to the statistical list-editor screen by pressing [2nd] [STAT] [F2]. Move the cursor to the top of the list. There should be a blank list and you should see ⓐ displayed in the upper right corner of the screen. You may have to press the right-arrow key. This indicates that the calculator is in ALPHABETIC mode. We need several more named lists. To name the first list, press L [ALPHA] [3] [ENTER] to name the list L3. If there is already a list named L3, clear the entries and go on to name the next list.

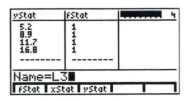

To name the next list L4, press ▶ L [ALPHA] [4] [ENTER]. If L4 already exists, clear its entries and go on to L5. To name the next list L5, press ▶ L [ALPHA] [5] [ENTER]. To name the next list L6, press ▶ L [ALPHA] [6] [ENTER].

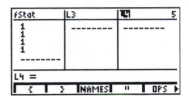 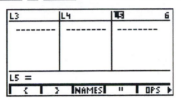 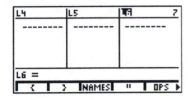

Use the left-arrow key and position the cursor over "L3." Then press [2nd] [LIST] [F3] and look at the menu at the bottom of the screen. Choose "xStat." You may have to press [MORE]. Press [–] [2nd] [STAT] [F5] [F1]. The keystrokes took us to the STATISTICAL VARIABLES menu where we selected "$\bar{x}$." Then press [ENTER].

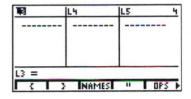 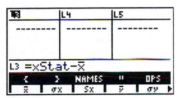 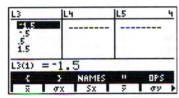

Now we have $c_i - \bar{x}$ in L3. To find $(c_i - \bar{x})^2$, move the cursor to the top of the L4 list. "L4" should be highlighted. Press [2nd] [LIST] [F3] and choose "L3" by pressing the appropriate F key. Then press [x^2] [ENTER].

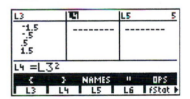

Now we have $(c_i - \bar{x})^2$ in L4. To find $\sum_{i=1}^{4}(c_i - \bar{x})^2 =$, go to the home screen by pressing [2nd] [QUIT]. Press

[2nd] [LIST] [F5] [MORE] [F1]. These keystrokes take us to the LIST OPERATIONS menu where we select "sum." Now press [2nd] [LIST] [F3] where we look on the bottom of the screen and select "L4" by pressing the appropriate F key.

Then press [ENTER]. Now we know $\sum_{i=1}^{4}(c_i - \bar{x})^2 = 5$.

 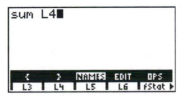 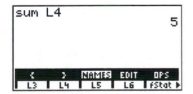

To find $d_i - \bar{y}$ return to the statistical list-editor screen by pressing [2nd] [STAT] [F2]. Move the cursor to the top of the L5 list. "L5" should be highlighted. Press [2nd] [LIST] [F3] and choose "yStat" by pressing the appropriate F key. You may have to press [MORE] to find it. Press [−] [2nd] [STAT] [F5] and press the appropriate F key to choose "$\bar{y}$." The keystrokes took us to the STATISTICAL VARIABLES menu where we selected "$\bar{y}$." Press [ENTER].

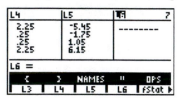

Now we have $d_i - \bar{y}$ in L5. To find $(c_i - \bar{x})(d_i - \bar{y})$, move the cursor to the top of the L6 list. "L6" should be highlighted. Press [2nd] [LIST] [F3]. Press the appropriate F key to select "L3." Press [×] and then press the appropriate F key to select "L5." Then press [ENTER].

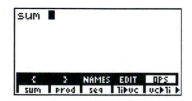

Now we have $(c_i - \bar{x})(d_i - \bar{y})$ in L6. To find $\sum_{i=1}^{4}(c_i - \bar{x})(d_i - \bar{y}) =$ go to the home screen by pressing [2nd] [QUIT]. Press [2nd] [LIST] [F5] [MORE] [F1]. These keystrokes take us to the LIST OPERATIONS menu where we select "sum." Now press [2nd] [LIST] [F3] where we look on the bottom of the screen and select "L6" by pressing the appropriate F key. Press [ENTER]. Now we know $\sum_{i=1}^{4}(c_i - \bar{x})(d_i - \bar{y}) = 18.8$.

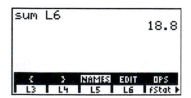

The slope is $\dfrac{18.8}{5} = 3.76$, thus the regression line is $y - 10.65 = 3.76(x - 2.5)$, which simplifies to $y = 3.76x + 1.25$.

Using the LINEAR REGRESSION feature is certainly quicker than this method, but both achieve the same result.

Below the original table is filled in.

Variable	c_i	d_i	$c_i - \overline{x}$	$(c_i - \overline{x})^2$	$d_i - \overline{y}$	$(c_i - \overline{x})(d_i - \overline{y})$
Calculator location	xStat	yStat	L3=xStat$-\overline{x}$	L4= L3^2	L5 = yStat$-\overline{y}$	L6 = L3 $\cdot$ L5
	1	5.2	-1.5	2.25	-5.45	8.175
	2	8.9	-0.5	0.25	-1.75	0.875
	3	11.7	0.5	0.25	1.05	0.525
	4	16.8	1.5	2.25	6.15	9.225
	$\sum_{i=1}^{4} c_i = 10$	$\sum_{i=1}^{4} d_i = 42.6$		$\sum_{i=1}^{4}(c_i - \overline{x})^2 = 5$		$\sum_{i=1}^{4}(c_i - \overline{x})(d_i - \overline{y}) =$ 18.8
	$\overline{x} = 2.5$	$\overline{y} = 10.65$				

Index
TI-83 Plus and TI-84 Plus Graphing Calculators